Information and Communication Technologies to Promote Social and Psychological Well-Being in the Air Force

A 2012 Survey of Airmen

Laura L. Miller, Laurie T. Martin, Douglas Yeung, Matthew D. Trujillo, Martha J. Timmer

RAND Project AIR FORCE

Prepared for the United States Air Force
Approved for public release; distribution unlimited

For more information on this publication, visit www.rand.org/t/RR695

Library of Congress Cataloging-in-Publication Data is available for this publication.
ISBN: 978-0-8330-8727-0

Published by the RAND Corporation, Santa Monica, Calif.

© Copyright 2014 RAND Corporation

RAND® is a registered trademark.

Preface

Little research exists to address Air Force concerns about the implications of social media, texting, video games, the Internet, and other information and communication technologies (ICTs) for the overall well-being of Airmen. Many of these means of electronic communication have been created or widely adopted only over the past decade, and thus scientific evidence of their impact on Americans' social support and mental health is still emerging. Additionally, American mental health care professionals have begun seeking ways to effectively and responsibly employ the use of email, video chat, instant messaging (IM), and other technological alternatives to supplement traditional face-to-face counseling sessions.

To help the Air Force understand the role of ICTs, including the Internet and social media, in Airman social support networks, mental health, and suicide prevention and the implications for outreach, RAND conducted a survey of Airmen in 2012. Participants consisted of 1,634 active-duty, 977 guard, and 868 reserve Airmen. The survey addressed the following research questions:

- What are the ICT usage rates among Airmen?
- Are social support and well-being related to ICT use?
- What are Airman attitudes about seeking and receiving health information and care via ICT?
- What do Airmen see as the overall impact of ICT on their well-being?
- How can the Air Force leverage ICT to promote the social and psychological well-being of Airmen?

This report provides highlights from the relevant scientific literature, including newly developing research; conveys the results of the 2012 survey; and offers recommendations for Air Force policy and practice.

The research reported here was commissioned by the Air Force Office of the Surgeon General (AF/SG) and conducted within the Manpower, Personnel, and Training and Program of RAND Project AIR FORCE.

RAND Project AIR FORCE

RAND Project AIR FORCE (PAF), a division of the RAND Corporation, is the U.S. Air Force's federally funded research and development center for studies and analysis. PAF provides the Air Force with independent analyses of policy alternatives affecting the development, employment, combat readiness, and support of current and future aerospace forces. Research is conducted in four programs: Force Modernization and Employment; Manpower, Personnel, and Training; Resource Management; and Strategy and Doctrine. The research reported here was prepared under contract FA7014-06-C-0001. Further information may be obtained from the Strategic Planning Division, Directorate of Plans, Hq USAF.

Additional information about PAF is available on our website:
http://www.rand.org/paf

Contents

Appendixes

Figures

Tables

Summary

In 2012, RAND conducted a pioneering exploratory survey of 3,479 active-duty, guard, and reserve Airmen on their use of information and communication technology (ICT), the association between ICT use and social and psychological well-being, and the potential for Air Force mental health professionals to leverage ICT to meet the needs of Airmen. In this study, ICT refers specifically to electronic-mediated communications such as texting, email, phone calls, instant messaging (IM), social media (e.g., Facebook, Twitter), video chat (e.g., Skype, video teleconferencing), blogging, online multiplayer video games and virtual worlds, and the Internet in general. Before we analyzed the results, the survey data were weighted to ensure that the analytic sample would be representative of the gender, age group, rank (officer, enlisted), and affiliation (active, guard, reserve) composition of the U.S. Air Force. We also reviewed the rapidly expanding literature on the use and impact of ICT. Although this is not a comparative study of American civilians and Airmen, the results of previous studies of civilians provided valuable background information and contributed to the development of the survey, as well as the interpretation of the results.

What Are the ICT Usage Rates Among Airmen?

Most Airmen report using some form of ICT every day for either work or nonwork purposes. Airmen are drawn from and live in a society where Internet access and cell-phone ownership are near universal among adults under the age of 50 (Pew Research Center, 2014). Among the Airmen in our survey, the forms of ICT most likely to be used every day for work purposes are email (almost 80 percent) and phone calls (over 60 percent). Twenty-eight percent reported using Internet sites for work purposes daily, and just over 10 percent text daily. Less than 10 percent use other ICT for work daily. The ICTs Airmen use most for nonwork purposes are texting (about 70 percent daily), email and phone calls (over 60 percent each), and social media and other Internet sites (about 45 percent each). About 15 percent of Airmen reported playing video games daily; 8 percent use IM, online chat rooms, and forums; 6 percent use video chat; and 3 percent use blogs every day.

We also collected self-reports on the number of hours each week Airmen spent using ICT. There is no standard definition of a "frequent user of ICT" in the research literature, but we found that approximately 10 percent of Airmen spend 20 or more hours a week using each of the most commonly used ICTs (texting, email, phone, social media, and video games and other Internet sites). We labeled these Airmen "frequent users" of each technology, relative to their fellow Air Force personnel. Nearly 30 percent of Airmen spend 20 or more hours a week using one or more ICTs.

Are Social Support and Well-Being Related to ICT Use?

ICT use is either positively or neutrally associated with social support and well-being, depending on the ICT. Prior research has demonstrated that social support networks are associated with resiliency and mental health. Data from our survey suggest that many Airmen use ICTs to augment interactions with real-world friends, family, and other Airmen, not replace them. For example, 65 percent of the Airmen surveyed use social media to keep in touch with friends and family, and almost half of them use it to keep in touch with other Airmen.

Moreover, 12 percent of the Airmen surveyed reported a high level of social support from people they communicate with *only* online. The minority of Airmen who do not use ICT such as social media, texting, email, or phone calls to keep in touch with others reported significantly worse mental health than did the overall survey sample. Perceived social support—both online and offline—was strongly and positively associated with self-reported mental health.

Relatively frequent ICT use was associated with measures of loneliness, but not with self-rated poor mental health or depressed mood. Although these associations were detected in our analyses, we cannot assume causality. It is possible that frequent users of social media, IM, video chat, and blogs are more likely to feel lonely than the study population overall because they are geographically separated from friends and family, not necessarily because their frequent ICT use led to social isolation.

Airmen who had witnessed cyberbullying were less likely to rate their own mental health as excellent (37 percent) than those who had not witnessed it (50 percent). We were interested in assessing cyberbullying, as it can lead to a number of poor mental health outcomes among victims, most notably the tragic outcome of suicide. We found that within the past year, one-quarter of Airmen reported being exposed to cyberbullying. Twenty-three percent of them witnessed someone else being bullied, and 7 percent were themselves the target. Although only 3 percent of all respondents admitted to perpetrating or attacking someone online, 13 percent reported intervening when they witnessed an attack.

In our survey, 6 percent of the Airmen met criteria for problematic Internet use on the Generalized Problematic Internet Use Scale 2. Psychologists have applied criteria from other impulse-control behaviors (such as gambling) to measure the degree to which excessive ICT use may be associated with poor well-being. Although most of the Airmen surveyed reported a small to moderate level of Internet use, a small minority may be struggling with a problematic level of Internet use that they are having difficulty controlling and that is causing trouble in their lives. This behavior could disrupt families, Airmen's units, and Airmen's abilities to perform their duties. The scale used in this survey captures only problematic Internet use: unknown is the percentage who may be experiencing a broader technology "addiction," which would include problematic use of mobile phones, computers, and video games in general.

In our sample, problematic Internet use was significantly correlated with self-rated poor mental health, depressed mood, and loneliness. Prior research suggests that Internet addiction may be a symptom of underlying problems rather than the root cause of them. Air Force leaders should be aware that time spent on cell phones, video games, or computers that results in neglect of duties, others, or self may not simply reflect a lack of self-discipline—it may be a sign of more serious mental health issues.

What Are Airman Attitudes About Seeking and Receiving Health Information and Care via Technology?

Airmen seek and use online mental health information, but they need help evaluating its quality. About one-third of the Airmen in our survey used ICT to learn about mental health topics or resources, including suicide prevention. Six percent had used ICT in the previous 12 months to learn about suicide prevention specifically, separate from Air Force training. Subsequently, 42 percent shared the mental health or suicide prevention information with friends or family; 20 percent used it to decide whether to see a mental health professional; and 18 percent reported that it changed the way they manage their mental health concerns. Additionally, one-quarter of the respondents used ICT to connect with others who have similar health concerns. Despite these findings, only about half of the survey respondents indicated on an eHealth literacy scale that they believe they can tell, locate, and discern high-quality from low-quality mental health resources on the Internet.

Airmen favor both ICT and face-to-face communication for learning about mental health from Air Force professionals. When asked about the ways that would be most helpful for receiving mental health outreach and educational or screening materials from the Air Force, two-thirds of the Airmen reported that they would like to receive such information via face-to-face interaction. But there was also interest in receiving such information through other means: 56 percent thought email would be helpful; 36 percent selected Internet sites; and 24 percent thought social media would be helpful.

Airmen favor ICT as a supplement or alternative to face-to-face communication with mental health professionals. Among the Airmen surveyed, 83 percent preferred face-to-face advice from a mental health professional; however, interest in other forms of communication was apparent. Email and phone calls were each endorsed by 38 percent of respondents as most helpful for seeking advice from a mental health care professional. For actual mental health care or treatment, 91 percent prefer face-to-face interaction, and 63 percent selected the in-person approach exclusively. Nevertheless, some respondents reported that it would be most helpful to receive mental health care or treatment through other communication means, including email, phone calls, social media and other Internet sites, text messaging, and video chat (e.g., Skype, video teleconferencing). Furthermore, 9 percent of Airmen surveyed would prefer mental health care or treatment delivered exclusively through ICT. It is important to note, however, that our survey respondents were not necessarily in need of or seeking treatment services. More research is needed to determine whether these preference patterns hold among individuals who are in need of or actively engaged with the mental health system.

What Do Airmen Believe Is the Overall Impact of ICT on Their Well-Being?

On the whole, most Airmen reported that, in their opinion, ICT has either a positive or neutral impact on their overall well-being and life satisfaction. Most Airmen (more than 80 percent) perceive a positive impact of text messaging, email, and phone calls; more than 60 percent reported the same about social media; and more than half believe that other Internet sites are beneficial. Video games, social media, blogs, and IM were most likely to be reported as having a negative impact, with about 5 percent of Airmen indicating such impacts.

Airmen believe that the ICTs most important for their social and mental well-being when they are away from their home station are phone calls (89 percent), email (82 percent), video chat (64 percent), and social media (62 percent). Similar patterns were found when Airmen were asked for self-reports about the impact of ICT during their most recent deployment: Almost none perceived a negative impact, and a majority indicated a positive impact of email, phone calls, video chat, and social media and other Internet sites.

Do ICT Use, Social Support, and Psychological Well-Being Differ Among Different Groups of Airmen?

There were few consistent demographic differences of statistical and practical significance in Airmen's survey responses about their ICT use and their well-being—the most numerous differences appeared in our comparisons of 18- to 24-year-old Airmen with other age groups while taking into account other demographic characteristics. Younger Airmen were more likely to score negatively on the Generalized Problematic Internet Use Scale 2 and to have had experiences with cyberbullying. They were also more likely than older Airmen to be frequent users of texting, video games, and social media. Moreover, Airmen ages 18 to 24 were more likely to report interest in some of the newer ICTs as modes for discussing mental health concerns and for receiving mental health outreach and educational or screening materials from the Air Force. Younger Airmen were more likely than older Airmen to report texting, social media, and IM as most important for their social and mental well-being when deployed or temporarily away from their home stations. Airmen ages 18 to 24 were also more likely than older Airmen to believe that each of the ICTs has a positive impact on their overall well-being and life satisfaction.

Other differences in survey responses were related to component, rank group, relationship status, having minor children, deployment/temporary duty (TDY) experience, and education level. For example, men, active-duty Airmen, and Airmen ages 18 to 24 were more likely to be frequent video game users (20 or more hours a week, among the top 10 percent of Airmen in terms of frequency). Frequent texting was more likely among women, Airmen ages 18 to 24, and divorced/widowed/separated Airmen; those same subgroups, along with never-married Airmen, were more likely to be frequent users of social media.

Active duty, enlisted Airmen, and women reported greater online social capital relative to their comparison groups. Guard, women, and 18- to 24-year-olds were more likely to report greater social capital among those they interact with in person. Enlisted Airmen, women, and never-married, divorced, separated, and widowed Airmen were more likely to self-rate their mental health negatively. Younger and/or never-married Airmen were more likely than other Airmen to report indicators of problematic Internet use. Younger, male, and/or enlisted Airmen were particularly likely to report experiences with cyberbullying. Some of these groups are more likely than others to perceive certain types of ICTs as being useful, helpful, or otherwise beneficial to their mental health. There were no consistently significant differences in survey responses among Airmen related to being a racial or ethnic minority.

How Can the Air Force Leverage ICT to Promote the Social and Psychological Well-Being of Airmen?

In conclusion, this study revealed the following:

- As in the general population, Airmen typically use ICTs to augment interactions with real-world friends, family, and other Airmen, not to replace them.
- Six percent of Airmen may be facing serious difficulties due to their poorly controlled and disruptive Internet use. Problematic Internet use was significantly correlated with self-rated poor mental health, depressed mood, and loneliness.
- Airmen seek and use online mental health information but need help evaluating its quality.
- Most Airmen prefer face-to-face communication with health professionals, but a sizable minority may prefer ICT alternatives.

The Air Force could help Airmen by

- Promoting policies that permit nondisruptive on-duty access to ICT to help strengthen family ties, social support networks, and well-being
- Enhancing strategic communication plans to increase Airman awareness of mental health resources, encourage responsible help-seeking, and decrease the stigma of mental health care
- Continuing to support access to ICTs during deployment
- Educating leaders that ICT "addiction" may be a sign of broader problems and encouraging appropriate referrals
- Educating Airmen about how to recognize signs of problematic ICT use and what resources are available to help them address it
- Teaching Airmen how to protect themselves against and respond to cyberbullying

- Providing basic guidelines for identifying credible websites and directing Airmen to credible sources through messaging and its own web pages
- Developing/leveraging alternative communication modes such as texting, social media, and Internet sites to supplement, but not replace, face-to-face interaction between Airmen and mental health care professionals
- Integrating into Air Force suicide prevention training information about the potential role of ICT for identifying suicide risk factors (e.g., via messages indicating intentions for self-harm) and Airman options for responding to suicidal messages, especially when Airmen do not personally know the individuals at risk or where they are located
- Exploring targeted mental health outreach through ICT during deployment
- Exploring through further research the experiences, needs, and attitudes of Airmen experiencing problematic ICT use
- Exploring through further research how Airmen in distress use ICT, to improve the Air Force's ability to detect and intervene when Airmen are at risk and thus potentially save lives
- Exploring through further research how Airmen would like to use ICT for mental health support, for what types of problems, and why
- Exploring ways to utilize ICTs to enhance mental health treatment adherence and symptom reduction.

Acknowledgments

We are grateful for the dedication of our project action officers: Col. Jay Stone, Deputy Director of Psychological Health and chair of the Air Force Integrated Delivery System; Maj. Michael McCarthy, Air Force Suicide Prevention Program Manager at the time of this study; and Lt. Col. Cherri Shireman, Chief of Operational Medicine at the Air Force Medical Support Agency. They were intellectually engaged throughout the study and also devoted considerable time to helping steer the survey application through the Air Force survey approval process. Insightful questions and suggestions from Col. John Forbes, Director of Psychological Health, helped to further develop our initial analyses of the survey results. Our reserve component liaisons were Lt. Col. David Bringhurst, Chief of the Behavioral Health Branch of the Air National Guard Readiness Center, and Lt. Col. David Ubelhor, Chief Mental Health Consultant to the Command Surgeon, Headquarters Air Force Reserve Command.

RAND colleague Rajeev Ramchand participated in the initial conceptualization of this project and offered feedback on our initial presentation of the literature review. Larry Hanser and Nelson Lim offered advice on our sampling strategy. Perry Firoz and Judith Mele supported this effort by pulling the survey sample from the Air Force personnel data extracts. Bas Weerman programmed the survey through RAND's MMIC™ (Multimode Interviewing Capability) information system, which hosted the online survey and provided a survey-response tracking tool. Tania Gutsche spent days at a time sending out large batches of survey invitations and survey reminders and monitored the survey help desk. Daniela Golinelli developed the survey weights for the analytic sample and wrote the description of that process. Coreen Farris coded the open-ended survey responses. Carl Rhodes, then-Director of the Manpower, Personnel, and Training and Program of RAND Project AIR FORCE, reviewed the project at multiple points; his questions and comments also helped improve our work. Susan Woodward prepared one of our figures and helped edit sections of text that appear in the summary. Robert Guffey greatly improved the formatting of the tables in Chapter Eight. We also thank Donna White, who assisted us with preparing this manuscript, particularly in creating many of the tables and formatting the bibliography. Jonathan Marten provided further support as the bibliography grew. Jane Siegel helped improve the appearance of the manuscript as well. Janet DeLand's editing skills made a significant contribution throughout the final report.

We would also like to acknowledge the contributions of Shelia Cotten at Michigan State University and Lynsay Ayer of RAND, who provided thorough and constructive reviews of an earlier version of our manuscript.

Abbreviations

ACA	American Counseling Association
AMA	American Medical Association
AMHCA	American Mental Health Counselors Association
APA	American Psychological Association
ATA	American Telemedicine Association
CPA	Canadian Psychological Association
DoD	U.S. Department of Defense
FSMB	Federation of State Medical Boards of the United States
GPIUS2	Generalized Problematic Internet Use Scale 2
ICT	information and communication technology
IM	instant messaging
NBCC	National Board of Certified Counselors
OPA	Ohio Psychological Association
PTSD	post-traumatic stress disorder
TBI	traumatic brain injury
TDY	temporary duty
VA	U.S. Department of Veterans Affairs
VTC	video teleconference

Chapter One

Introduction

Little research addresses Air Force leadership's concerns that the increasing use of social media, texting, video games, the Internet, and other information and communication technologies (ICTs) could be degrading Airman social skills, support networks, and overall well-being. The Air Force Surgeon General's office sponsored the research reported here to learn how ICT use relates to Airman social and psychological well-being. In this study, ICT refers specifically to electronic-mediated communications, such as texting, email, phone calls, instant messaging (IM), social media (e.g., Facebook, Twitter), video chat (e.g., Skype, video teleconferencing), blogging,[1] online multiplayer video games and virtual worlds, and the Internet in general. The Air Force Surgeon General also asked RAND to explore whether there is potential value in using these ICTs to promote resilience, meet the need for mental health care, and prevent suicide. To address this knowledge gap, we reviewed the growing but relatively new body of research in this field and conducted an online survey of Airmen in 2012. Survey respondents comprised 1,634 active-duty, 977 guard, and 868 reserve Airmen, for a total of 3,479 Airmen. The survey addressed the following research questions:

- What are the ICT usage rates among Airmen?
- Are social support and well-being related to ICT use?
- What are Airman attitudes about seeking and receiving health information and care via ICT?
- What do Airmen see as the overall impact of ICT on their well-being?
- How can the Air Force leverage ICT to promote the social and psychological well-being of Airmen?

This report reviews emerging research on these topics, provides analyses of the results of the 2012 survey of Airmen, and offers recommendations for Air Force policy and practice.

Organization of This Report

Chapter Two describes the survey methodology—how the survey instrument was developed, how Airmen were sampled to participate, who participated in the survey, and how the responses were weighted to match the demographic composition of the Air Force to create the analytic sample used for our analyses. Chapter Three offers some background on ICT use in the United States and provides the ICT usage rates reported by Airmen in the survey. Chapter Four conveys

---

[1] *Blogging* and *blogs* are abbreviations for weblogging and weblogs, respectively.

how ICTs relate to Airman social support networks. Chapter Five delves into the association between Airman ICT use and mental health. Chapter Six reports on Airmen's use of ICT to learn about mental health care and their level of interest in the Air Force using ICT to provide mental health information, advice, care, and treatment. Chapter Seven reports Airmen's perceptions of the impact of ICTs on their overall social and mental well-being. Chapter Eight explores demographic differences in Airman survey responses that could potentially inform Air Force outreach efforts. Chapter Nine ties the findings of the previous chapters together and presents recommendations that stem from the research. The survey instrument is reproduced in Appendix A, and the survey invitation is shown in Appendix B. Technical details about the sampling and weighting methods used are provided in Appendix C. Further resources related to mental health promotion and intervention are provided in Appendixes D, E, and F.

Chapter Two
Study Methods

This chapter provides a summary of the study methods we used, including the development of the survey instrument, design of the sampling frame, and administration of the survey. It also provides information on response rates and how the sample was weighted to be representative of key demographics of the Air Force military population as a whole.[1]

Development of the Survey Instrument

Input from Air Force representatives and the scientific literature contributed to the development of the survey instrument. We first met with the project monitors to learn about the concerns of Air Force leaders and the types of information that would be valuable given their questions of interest. Priority areas of interest included ICT use and overuse, mental health and well-being, social support, information-seeking related to mental health, and interactions with mental health providers. These topics were prioritized for inclusion in the survey. We then conducted a review of the existing literature on selected ICTs and their role in promoting or harming social networks, mental health, and suicide prevention. Insights from this literature review informed the development of the survey and are presented in the subsequent chapters along with the associated survey results. We provide information from other surveys and studies as context for this study, but we do not attempt to directly compare U.S. civilian and Air Force populations.

We developed an initial draft of the survey that reflected the interests of the research sponsor and the guiding research questions. We included published scales developed by other researchers that have been demonstrated to measure their constructs reliably. Those scales and the sources documenting their development and testing are the online and offline bonding scales from the Internet Social Capital Scales (Williams, 2006), the Three-Item Loneliness Scale (Hughes, Waite, Hawkley, and Cacioppo, 2004), the Generalized Problematic Internet Use Scale 2 (GPIUS2) (Caplan 2010), and the Patient Health Questionnaire 2 (Kroenke, Spitzer, and Williams, 2003). We also adapted an eHealth literacy scale to refer specifically to mental health resources rather than health resources more generally (Norman and Skinner, 2006a). The project action officers from the Air Force Surgeon General's officer reviewed the draft instrument and

---

[1] RAND's Institutional Review Board approved the conduct of this research. RAND also received the required approvals from the Air Force Manpower Agency, which issues the Air Force survey control numbers; from the Air Force Research Oversight and Compliance Office, whose review functioned as a second Institutional Review Board review; and from the Chief Information Office, which granted a waiver for us to host an Air Force survey on a nonmilitary website.

provided feedback on the items. The researchers then met with a group of enlisted Airmen at Andrews Air Force Base, who also reviewed the survey draft and made recommendations concerning both the front matter of the survey (e.g., the introductory language) and individual items. Ultimately, the research team prioritized the survey items and trimmed those less central to the line of inquiry in order to reduce the expected time for respondents to complete the survey to about 15–20 minutes.

As a safety precaution, information about how Airmen in emotional crisis could receive confidential help appeared at the bottom of each survey screen that contained questions related to mental or emotional health and at the introduction and conclusion of the survey.

The survey tool, including the informed consent statement and these referral messages, is reproduced in Appendix A.

Survey Administration

Every active-duty, guard, and reserve Airman age 18 and older was eligible to be selected for this survey. The sampling method sought to assure sufficient numbers of respondents to analyze rank group, gender, and age group within each of the active-duty, guard, and reserve populations but to also minimize the number of Airmen who would be asked to participate in the survey. In total, email invitations were sent to 9,437 active-duty Airmen, 9,000 guardsmen, and 9,000 reserve Airmen. The template for the survey invitations is shown in Appendix B. Details about the sampling design are provided in Appendix C. The first survey invitations were emailed on June 25, 2012, personally addressed to each invitee. The survey closed on August 25, 2012, for active-duty Airmen and October 30, 2012, for guard and reserve Airmen.[2]

The survey was anonymous, meaning that there was no way for the researchers or anyone else to connect survey responses to specific individuals. An algorithm generated a unique link to the survey from each email invitation that could not be retrieved even by the computer programmers who created the algorithm. Since we did not know which addressees had and had not completed a survey, those who did not elect to opt out of the survey were sent reminder emails—approximately every two weeks for active-duty invitees and every month for reservists. Respondents were encouraged to create their own survey access passwords so that they could return to complete the survey through the same link if they did not finish in a single session. However, after a lapse of three days from their prior session, they would no longer have access to previously entered survey items (they could access only the items they had not yet completed), which provided an added level of security to their responses. Those who forgot their passwords could contact RAND to be sent a new survey link.

---

[2] The reservists were allotted a greater window for participation because, overall, they have less frequent access to Air Force email, particularly during the summer months.

4

The online survey was hosted on a RAND Corporation website ending in "rand.org." Since the survey was not hosted on a military website, participants did not have to respond to the survey from their military computer and/or log in with their military identity card. This offered them greater flexibility in when and where they could participate. For those who wanted to verify that the survey invitation was for a legitimate study, the invitation included the Air Force survey control number issued by the Air Force Survey Office and a link to a letter of endorsement from the Air Force Deputy Surgeon General.

Survey participation was voluntary, and the respondents had the option to stop taking the survey at any time or to skip items and still advance through the survey. Partially completed surveys (those that were started but not completed) were omitted, as were those that did not include the basic demographic information needed to weight the sample proportionately. Responses of participants who skipped only individual survey items were included in the analytic sample even though they may not have been usable in every analysis. An inspection of skipped items revealed no patterns; that is, we did not observe a pattern of Airmen tending to skip potentially sensitive questions, such as those on self-rated mental health or depression.

Respondents and the Weighted Analytic Sample

A total of 3,479 Airmen responded to the survey—1,634 active-duty, 977 guard, and 868 reserve Airmen. This sample size was sufficient for the planned data analyses. It compares favorably with surveys used to provide estimates for the much larger entire U.S. adult population: Gallup's public opinion polls typically have 1,000 respondents, and the General Social Survey sponsored by the National Science Foundation since 1972 has approximately 1,500 participants.[3]

We had planned for a survey launch in April and anticipated a response rate close to 30 percent; however, due to delays beyond our control we were unable to administer the survey until the summer months, a period in which it is more challenging to reach Airmen, for a number of reasons. It is common for Airmen to have to move from one base to the next during the summer months. In some cases, a reassignment to a new duty station can also mean that the Air Force email addresses would change from those we had on record for survey invitations. Summer, when children are out of school, is also the period in which family vacations are more common. Also guard and reserve Airmen may be more difficult to reach in the summer because some units do not meet during that period; therefore, we extended the survey period for these invitees. We did experience the lower response rates we had expected: Response rates for active, guard, and reserve Airmen were 17 percent, 11 percent, and 10 percent of the invitees,

---

[3] For more information on Gallup polls, see Gallup (undated). To learn more about the National Opinion Research Center's General Social Science survey, see http://www3.norc.org/GSS+Website/.

respectively, and reflect both the original invitees and an additional set of Airmen invited to participate during a much shorter window of time toward the end of the survey period.

Calculations first revealed that a simple random sample of Airmen would recruit far more active-duty Airmen than needed to meet the desired number of guard and reserve Airmen in each of the gender, age group, and rank categories, so we drew a random sample of 4,500 Airmen from each these populations, with the intention of weighting the survey responses to reflect the actual proportion in the Air Force (which in 2012 was 65 percent active-duty, 21 percent guard, and 14 percent reserve, rather than 33 percent each). This strategy would produce enough respondents for every subgroup except active-duty Airmen ages 45 and above. Rather than greatly increase the number of invitations just to acquire enough Airmen in that one subgroup, we strategically oversampled that population (an additional 437 invitations were sent to them). Once it became clear (in July) that the overall response rates were going to be lower than anticipated, we sent out a new round of invitations to an additional sample of Airmen in early August, doubling our sample size, although it was not necessary to oversample the older active-duty age group again.

Because of the sampling design and the differential response patterns among subgroups, it was necessary to build post-stratification weights so that the proportions of respondents in our analytic sample matched those of the entire Air Force population. For example, although we had been concerned about obtaining enough active-duty participants ages 45 and older, that subgroup, and indeed older respondents in general, were disproportionately likely to respond to the survey, and the youngest category of Airmen was disproportionately less likely to respond. Additional research would be necessary to understand these response rates, but by weighting the data, we ensured that the demographic subgroups that were known to be more likely to respond are not overrepresented in our analyses. For example, Airmen ages 18 to 24 are 27 percent of our analytic sample because they constituted 27 percent of the Air Force population at the time. Technical details about the methods used to weight the sample are provided in Appendix C. The weights for each group were based on the selected demographics (gender, age group, rank group, and active, guard, or reserve affiliation), along with frequencies drawn from the active-duty sample. To ensure consistency in data analysis, we constructed a weighted dataset using these weights, with which we conducted all of our analyses.

Data Analysis

Throughout Chapters Three through Seven, we report the results of univariate analyses for the overall weighted analytic sample, along with the unweighted number (n) of respondents for each item or scale. We also utilized multivariate analysis to determine whether there were statistically ($p < 0.01$) and practically significant demographic differences among Airman survey responses. These analyses are described more fully in Chapter Eight, where we note key differences, although for the most part, Airman responses were quite similar.

Limitations

This study was undertaken to explore Airman ICT use, social support, and well-being as well as potential Air Force use of ICT to bolster its mental health care system. By necessity, this project was bounded to include only the highest-priority research questions, although many other important questions remain.

In using the survey method over interviews or focus groups and by choosing breadth over depth, we were able to improve the generalizability of our results and provide the Air Force an overview of Airman experiences on a wide array of topics. The sample size, however, limits the degree to which some subpopulation differences in the data can be mined. The lower response rate among the youngest Airmen is also of concern. Although we employed the standard practice of weighting the survey results to correct for variation in subgroup responses, without an accompanying nonresponse bias study we cannot be certain that those who did not respond would have provided answers similar to those of the Airmen who did respond. A nonresponse bias study could also be designed to detect whether the results of a summer survey are statistically and practically different from survey results administered during other times of the year.

Future studies will need to provide additional depth on the topics in the survey, including

- causality
- demographic and military Service differences
- rich contextual information surrounding Airman attitudes and experiences
- the impact of Airmen's ICT use on their units
- the impact of Airmen's ICT use on their families.

As an illustrative example, we describe some limitations of the study's finding of an association between frequent ICT use and measures of loneliness (Chapter Five). We do not know whether frequent ICT use led to isolation and loneliness, or whether isolation and loneliness led to an increased level of ICT use. We also do not know whether ICT use helped alleviate those feelings of loneliness, perhaps as Airmen settled in to a new base environment or managed separation from friends and family. Moreover, we cannot say why or how Airmen use ICT when they are lonely: Are they trying to meet new people, including people to date? Are they reaching out to people they already know or trying to connect with people they used to know? Are they trying to distract themselves or cheer themselves up with entertainment? Are they shopping compulsively?

We also do not know the degree to which the results may have been influenced by the survey mode. Airmen were sent an email inviting them to participate in the survey, and a direct link to the survey was embedded in the email. Given that this survey was about ICT and that it was conducted online, we expected that our results might be most likely to represent Airmen who are more active Internet or ICT users. However, we questioned that assumption after it was apparent that the youngest age group (18- to 24-year-olds) was least likely to participate in the survey (see

Table 2.1 in Chapter Two). Researchers may wish to examine in future efforts multiple survey modes (e.g., paper, telephone, online, mobile app) to determine whether response rates and responses vary by mode used.

Conclusion

Through a survey of Air Force military personnel, we sought to learn more about Airmen's use of ICTs and the implications for their well-being and Air Force mental health care services. The survey instrument was developed in coordination with the research sponsor and reflected a review of the scientific literature.

Survey participation was truly anonymous—no one, not even the researchers, was able to link individuals to their survey responses. Participation was also voluntary. Invitations were sent out over the summer of 2012 for active-duty members and continued through the end of October 2012 for guard and reserve Airmen.

The analytic sample includes only the completed surveys that contained the essential demographic characteristics necessary to weight the responses, to ensure that the results would not be skewed by a disproportionate response from particular subgroups. The findings presented in the remaining chapters were calculated using the analytic weighted sample. This study provides an exploratory examination of Airman beliefs and experiences related to ICT use, social support, and psychological well-being.

Chapter Three

Frequency of ICT Use Among Airmen

The past decade has seen an explosion in the use of mobile technology devices, such as laptops, mobile phones, and computer tablets, and an ever-increasing array of application software ("apps") expands the communication capabilities of those devices. Skype gives users the ability to make phone calls and video chat/video teleconference (VTC) from their computers and handheld devices. Other applications provide "places" such as Facebook and Second Life where people can meet and interact in cyberspace. Applications can increasingly be accessed from multiple types of devices, such as smartphones, video game consoles, and computer tablets. To select ICTs from among the vast array of ICT applications and devices for inclusion in this study, we reviewed technology usage in the U.S. population, considered the ICTs that had been explored thus far in scientific research, and solicited the perspectives of our research sponsor. We pursued information about Airman use of

- phone calls
- email
- text messaging
- video chat (Skype, VTC)
- social networking sites, such as Facebook, Twitter, Google Plus, and LinkedIn
- video games, including virtual worlds and social games, such as Small World, Second Life, World of Warcraft
- instant messaging, online chat rooms and forums
- blogs
- other Internet sites.[1]

ICTs not specifically named or described in the study include e-book readers such as Kindle and Nook; geolocation services, such as Foursquare, that can broadcast a user's location to others; online dating services, such as Match.com and eHarmony; music services, such as Spotify, where users can build and share playlists with one another; picture-sharing services, such as Instagram; and video-sharing sites, such as YouTube. Some of these ICTs, however, could fall into the "social media" or "other Internet sites" categories.

This chapter first provides the context of ICT usage in the United States and then reports the findings from our survey.

---

[1] This list reflects the wording used in the survey to describe these ICTs.

The Context of ICT Usage Among American Adults

Although Airmen may differ from the American population in ways that could relate to ICT use (e.g., younger average age, employment status), societal patterns of ICT use provide context for this study. In this section, we present recent statistics on ICT use, highlighting where possible results reported for American adults under the age of 50 (the age range of more than 90 percent of Airmen).

The Internet is now a widely available source of information and social interaction. Since 1995, the Pew Research Center's Internet & American Life project has been documenting the rapid growth of the Internet and mobile connectivity among adults in the United States by conducting nationwide random phone surveys, online surveys, and qualitative research. In early 2014, Pew sponsored a telephone poll (cell and landline) of a nationally representative sample of 1,006 American adults living in the continental United States and weighted the results to correct for known demographic discrepancies. Pew found that 89 percent of 18- to 29-year-olds and 86 percent of 30- to 49-year-olds are computer users (Pew Research Center, 2014, p. 12). While Internet access was once restricted to desktop computers, the Internet can now be accessed on devices including laptops, computer tablets, televisions, video game consoles, and mobile phones. Two-thirds of all the adult cell-phone owners in the Pew survey say they use their phone to access the Internet, and one-third say their phone is their primary device for Internet access (Pew Research Center, 2014, p. 13). Accordingly, Internet use is extremely high—97 percent of 18- to 29-year-olds and 93 percent of 30- to 49-year-olds reported being Internet users (Pew Research Center, 2014, p. 18).

Telephone landlines are increasingly disappearing in favor of the exclusive use of cell phones for calls (known as *wireless substitution*), especially among young adults. In 2013, approximately 54 percent of Americans ages 18 to 24, 66 percent of those ages 25 to 29, 60 percent of those ages 30 to 34, and 45 percent of those ages 35 to 44 lived in wireless-only households, compared with only 30 percent of adults ages 45 to 64 and 13 percent of adults 65 and over (Blumberg and Luke, 2013, p. 3). Basic cell phones provide voice calls and texting, and some provide limited Internet access. Nearly all American adults under the age of 50 are cell-phone owners (98 percent of 18- to 29-year-olds and 97 percent of 30- to 49-year-olds [Pew Research Center, 2014, p. 14]). Smartphones can offer access to Internet sites on a level comparable to that of personal computers, along with various other means of communications (e.g., IM, video chat) and mobile applications—downloadable programs that offer specific services. Smartphones are particularly popular among Americans under the age of 50: 83 percent of 18- to 29-year-olds and 74 percent of 30- to 49-year-olds have them (Pew Research Center, 2014, p. 16). The mobile apps accessible from smartphones have themselves gained significant popularity. Flurry (a mobile analytics firm) reports that in 2011, more Internet time was spent on mobile apps than on both desktop computers and mobile devices (Newark-French, 2012).

Skype enables free or low-cost phone calls and video chat for those with computers or smartphones who use the Internet to connect to anyone else on Skype. The ability to see and hear loved ones at no cost is an invaluable resource, particularly for military personnel serving overseas. Apple's FaceTime is a similar application for callers using compatible systems and models of its computers or mobile electronic devices.

Instant messaging, online chat rooms, and forums and blogs are other ways Internet users can connect with others, discuss interests, and participate in interest-driven communities. Blogs are websites with a distinctive reverse chronological format that allow users to create personalized online content similar to that of diaries or journals. While blogging's popularity increased between 2008 and 2010 among online adults ages 34 to 35 (from 10 percent to 16 percent), blogging among younger adults (ages 18 to 33) declined from 20 percent to 18 percent, although younger adults do bloglike activities through other online formats, such as posting updates about their lives on social media (Zickuhr, 2010, p. 21)

Social networking sites (also known as social networking services or social media) include widely used sites, such as Facebook, LinkedIn, Myspace, Twitter, Pinterest, and Instagram, that allow users to connect with others, share pictures or other information, and play games or otherwise interact. The popularity of social media has increased rapidly. The proportion of online individuals who used it more than doubled between 2008 (29 percent) and 2011 (65 percent) (Madden and Zickuhr, 2011). In addition to this increase in overall social media usage, the frequency of visits and the likelihood of visiting social media sites in a typical day have also increased (Lenhart, 2009).

In 2013, Pew found that 73 percent of American adults who use the Internet visit a social media site (Duggan and Smith, 2013, p. 1). Facebook is by far the dominant platform, used by 84 percent of 18- to 29-year-olds and 79 percent of 30- to 49-year-olds (Duggan and Smith, 2013, p. 4). The primary reasons users cite for participating in social media include communicating with existing offline friends, family, and other connections and reconnecting with old friends (Hampton et al., 2011; Lenhart, 2009). Previous research has found that very few Facebook friends are online-only friends (i.e., people whom users have never met in person) (Hampton et al., 2011).

Video games can be communication technologies as well and are available through a number of conduits, including on social media and other online sites; dedicated gaming consoles (e.g., Microsoft Xbox, Sony PlayStation 3, Nintendo Wii); personal computers; or mobile devices. According to a 2013 gaming-industry survey of more than 2,000 nationally representative households, 62 percent of gamers play games with others, either in person or online (Entertainment Software Association, 2013, p. 5). Massive multiplayer online games and online role-playing games, such as World of Warcraft, offer particularly engaging and interactive experiences. Virtual worlds, such as Second Life, also allow game players to explore an often extremely detailed, extensive, realistic world and interact with other players in ways that may resemble the real world very closely. Relative to overall gaming, mobile gaming is growing in size and importance. Mobile gaming also extends beyond just cell phones—84 percent of tablet owners play games (Geekaphone, 2011). Geekaphone (a service that enables users to compare cell phones) predicts that the mobile game industry will increase from $8 billion in 2011 to $54 billion by 2015. Further, according to Flurry, most of people's time on mobile apps is spent playing games (Newark-French, 2012).

Air Force leaders' concerns about the rise of mobile phone and Internet use among American adults, particularly among young adults, and possible implications for Airman social and psychological well-being contributed to the decision to the fund the research described in this report.

ICTs Most Frequently Used by Airmen for Work and Nonwork Purposes

The first goal of our survey was to determine the extent to which Airmen use various forms of ICT and for what purposes. Airman access to ICTs during duty hours could vary, depending on their job responsibilities and location. Some Airmen might have routine access to computers and the Internet and may even have been issued smartphones for work purposes. Others may work on computers but in secure environments with very limited access to the Internet or even their personal cell phones during work hours. Still others may have little to no access to ICTs at all during the workday. To avoid conflating ICT use that might be a part of Airman duties from nonprofessional use, we asked survey respondents to report how many times a week they used each ICT in the past 30 days for *work* and *nonwork* purposes separately.[2] Figure 3.1 shows the frequency with which Airmen used each ICT for nonwork purposes.

ICT usage rates varied widely. Text messaging, email, and phone calls were the ICTs most frequently used, i.e., the greatest percentage of Airmen (more than 60 percent) reported using them every day in the past 30 days. In contrast, blogs were the least frequently used, with more than 70 percent of respondents not using them at all in the past 30 days.

---

[2] Respondents were asked to reflect upon the previous four weeks, and thus their responses may reflect some recall bias. We opted for the longer recall time frame because we were seeking the Airmen's sense of their typical use and were concerned that a seven-day time frame might not adequately capture it given the many potential disruptions (vacations, geographic relocations, temporary duty [TDY] assignments, flying missions to overseas bases, etc.), particularly in the summer when the survey was administered.

Figure 3.1. Airman ICT Usage for Nonwork Purposes in the Previous 30 Days

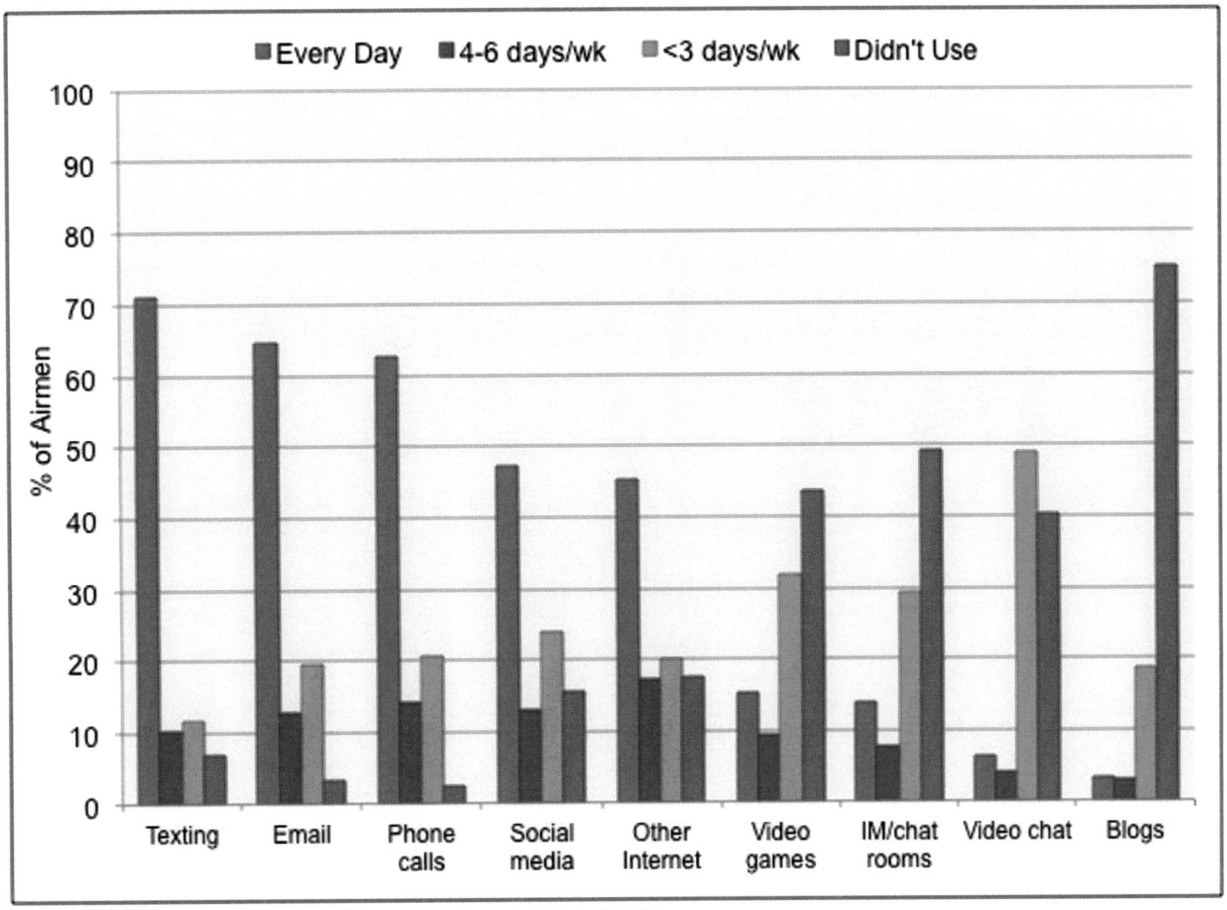

NOTE: Number of respondents (N) ranges from 3,426 to 3,469.

Airmen used the ICTs listed in the survey for work purposes far less frequently than they used them for nonwork purposes. As shown in Figure 3.2, email and phone calls were the ICTs most frequently used for work, with almost 80 percent of respondents using email and more than 60 percent using phone calls daily during the previous 30 days.

Figure 3.2. Airman ICT Usage as a Part of Their Job in the Previous 30 Days

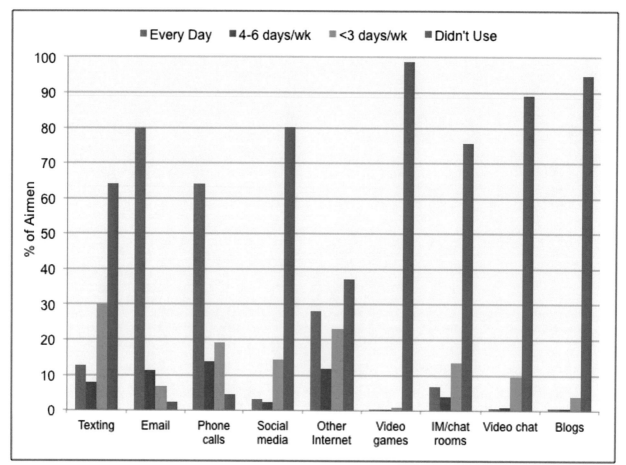

NOTE: N ranges from 3,452 to 3,462.

Relatively Frequent ICT Use Among Airmen for Nonwork Purposes

In addition to asking Airmen how many days a week in the past 30 days they used ICTs, the survey asked how much time they spent using each ICT on those days when they did use them. Combining the responses to those two questions, we categorized weekly levels of use for each ICT for nonwork purposes. To evaluate Airmen who are frequent users of these ICTs relative to other Airmen, we created a category of "frequent users." There is no research standard for "frequent use" of ICTs, and with their ever-growing availability, even if a standard had been developed just a few years ago, it might no longer be applicable today. An examination of the distribution of data by ICT did not suggest any natural break point for how best to define "frequent use." We therefore defined frequent ICT use as 20 or more hours per week for nonwork purposes, which is approximately the top tenth percentile of Airmen for the most frequently used ICTs. Again, our purpose was simply to denote what constitutes the greatest use among Airmen, not to suggest that 20 hours or more a week is excessive use or that Airmen are

14

more frequent users than their civilian counterparts. Figure 3.3 displays the percentage of Airmen who are frequent technology users for each type of ICT (for nonwork purposes only).

Figure 3.3. Airman Frequent ICT Usage for Nonwork Purposes in the Previous 30 Days

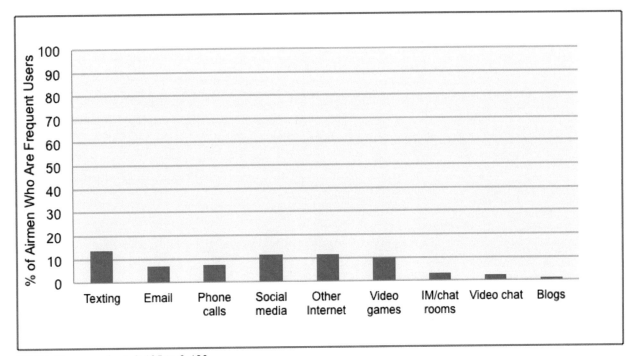

NOTE: N ranges from 3,405 to 3,463.

Airmen can use more than one ICT simultaneously—they can talk on the phone while playing video games, or text their friends about information while they are reading it online. We found that 29 percent of Airmen use at least one of the ICTs listed in the survey 20 hours or more a week for nonwork purposes.

Conclusion

The many ICTs available to today's Airmen are widely used both in the general U.S. adult population and among the Airmen in our survey. Our survey respondents indicated that text messaging, email, phone calls, social media, and other Internet sites were the ICTs most frequently used for nonwork purposes, and email and phone calls were the ones most frequently used as a part of their jobs. We found that for each of the most commonly used ICTs and for video games, approximately 10 percent of Airmen spent 20 or more hours per week online in the past 30 days for nonwork purposes. We classified that 10 percent as frequent users relative to other Airmen in our sample. Overall, 29 percent of the respondents use one or more ICTs for nonwork purposes 20 hours or more a week.

Chapter Four

Airman ICT Use and Social Support

ICTs can be an enabler for social interaction, providing a low-cost platform for sharing remotely (Shirky, 2010). People are intrinsically motivated to participate meaningfully in social communities (McGene, 2013; Ryan and Deci 2000). Positive social interactions, social support, and a sense of belonging can help individuals cope with stressful life events and promote psychological well-being (McGene, 2013). Moreover, meta analyses have demonstrated that lack of social support during and after trauma can put trauma-exposed individuals at higher risk for post-traumatic stress disorder (PTSD) (Brewin, Andrews and Valentine, 2000; Ozer et al., 2008). Specific to the military, unit cohesion—measured by a scale assessing whether unit members "cooperate with each other, know they can depend on each other, and stand up for each other"— has been found to be a significant protective factor against suicidal ideation among post-deployment Army combat soldiers (Mitchell et al., 2012, p. 488). Similarly, a study of United Kingdom Armed Forces deployed to Afghanistan found that greater levels of unit cohesion, along with greater morale and good leadership, were associated with lower rates of both PTSD and common mental disorders (Jones et al., 2012).

This chapter examines Airmen's use of ICT to keep in touch with other Airmen and friends and family; to meet new people; to connect with others they know only online; and to connect with others who may share a similar health concern. Finally, it evaluates the degree to which survey respondents experience social support from those they know primarily or exclusively online, as well as support from those they know in person. Negative aspects of online activity, including cyberbullying and "addictive" behavior, are discussed in Chapter Five.

Social Relationships Can Benefit from ICT

Historically, people have met and interacted with others who are in the same location at the same time. ICT has facilitated a unique historical situation, however, where common interests rather than place and time can become the primary drivers of human connection (Baym, 2010). Creating and sharing content online can provide people with a sense of efficacy, allow them to feel that they are needed, and increase prestige and status within their group (Baym, 2010). Sharing personally created content increased among adults between 2007 and 2009 (Lenhart et al., 2010). This may reflect simple shifts in many activities from in-person to online interactions, but it may also represent increased perceptions of the value of such activities. Teenagers who engage in "expressive" use of the Internet (e.g., reading and writing blogs, communicating with others) are more likely to also use social media, which afford similar benefits (Tufekci, 2008).

Much online content creation by teenagers is motivated by individuals' desire to document their lives to share with family and friends (Ito et al., 2010). Social media allows users to write status updates and share pictures and videos. Mobile phones provide constant Internet access as well as location-based services (e.g., sharing one's location). Internet users can gain status and credibility within a specific community by creating online guides to those worlds to help other users navigate (Ito et al., 2010). Blogs, forums, and chat rooms allow people to discuss common interests or write reviews of restaurants, music, events, products, and movies. Video games, particularly multiplayer online games and virtual worlds, provide opportunities to collaborate with fellow gamers and to create materials (e.g., game maps, strategy guides, fan fiction) for the gaming community. Below we highlight findings on relationships between ICT use and social support from the literature and from our survey of Airmen.

Maintaining Existing Relationships Is a Primary Reason for Using ICT

ICTs appear to complement existing means for maintaining social relationships, and there is evidence that ICTs can facilitate the formation of new relationships. For example, one laboratory experiment showed that people liked each other better when they first met over the Internet rather than face-to-face (McKenna, Green, and Gleason, 2002). However, people generally use online services to increase contact with existing friends rather than only to form new relationships (Fischer, 2011; McKenna, Green, and Gleason, 2002). Social media is used most frequently for maintaining existing relationships, allowing people to communicate indirectly, e.g., by photo sharing (Baym, 2010). Baym also notes that most friends do not rely exclusively on social media to communicate with each other; those who use more ICT and media are in fact more likely than others to communicate face-to-face as well and to maintain social relationships in general. A 2011 Nielsen study found that the top reason cited for adding someone as a friend on Facebook was knowing that person in real life (Nielsen, 2011). A 2014 Pew survey found that 67 percent of Internet users believe that their relationships with their friends and family have been strengthened through online communication, while 18 percent believe their relationships have been weakened (Pew Research Center, 2014, p. 24).

People's ICT networks, such as their social media networks, may be based on their existing real-life networks. For example, a nationally representative Pew study of social networking sites found that people with large social networks "gravitate to specific social networking platforms, such as LinkedIn and Twitter. The size of their overall networks is no larger (or smaller) than what we would expect given their existing characteristics and propensities" (Hampton et al., 2011, p. 42). That study also noted that "frequent use of Facebook is associated with having more overall close ties" and that Facebook users are more trusting, have more close relationships, and receive more social support than the average American (Hampton et al., 2011, p. 42). Greater cell-phone and chat use were also associated with having a larger overall network.

ICT users appear not only to possess more social connections but also to pay more attention to attending to and supporting them. Among a sample of college students, social media users were more interested than nonusers in "social grooming" activities, such as keeping in touch with and expressing curiosity about others (Tufekci, 2008).

Airmen Use ICT to Maintain Real-World Ties and to Make Online Contacts

As demonstrated in previous research on broad populations, the relationship between ICT use and social relationships may be either positive or negative. Echoing many similar concerns expressed in the general public, Air Force leaders have been concerned about the potentially isolating effects of ICT use. We therefore asked Airmen about whether they used various forms of ICT to maintain relationships with family, friends, and fellow Airmen.

Most of the Airmen in our survey use ICT to maintain real-world ties; less than 10 percent reported not using any ICTs to keep in touch with friends, family, and other Airmen or to make plans with anyone. As shown in Table 4.1, texting, email, phone calls, and social media are the most widely used ICTs for these purposes, and 37 percent of the respondents use video chat to keep in touch with friends and family.

Table 4.1. Airman ICT Usage for Maintaining Real-World Ties

ICT	Keep in Touch with Other Airmen (percent)	Keep in Touch with Friends and Family (percent)	Make Plans for in-Person Activities with People (percent)
Texting	59	77	75
Email	75	72	57
Phone calls	63	90	83
Social media	47	65	40
Other Internet sites	4	6	3
Video games	9	14	4
IM/chat rooms	15	17	11
Video chat	5	37	5
Blogs	1	3	1
Did not do this	6	3	6

NOTE: N ranges from 3,455 to 3,458.

Airmen also use ICTs to meet and interact with new contacts, although at rates lower than their rates for communicating with people they already know. Still, approximately 40 percent reported meeting new people via ICTs, 60 percent connect with people they know only online, and 35 percent reported using ICTs to connect with others with a similar health concern (Table 4.2).

Table 4.2. Airman ICT Usage for Other Social Networking Purposes

ICT	Meet New People (percent)	Connect with People Known Only Online (percent)	Connect with Others with a Similar Health Concern (percent)
Texting	8	11	5
Email	12	22	9
Phone calls	11	6	8
Social media	30	42	12
Other Internet sites	7	6	25
Video games	9	13	0
IM/chat rooms	7	12	6
Video chat	1	4	1
Blogs	2	3	5
Did not do this	57	42	65

NOTE: N ranges from 3,432 to 3,447.

Airmen Reported Greater Social Support from People They Know in Person Than from People They Know Only Online

Assessing Airman online social support was a key goal of this research, along with addressing the question of whether increasing ICT use and online interaction has led to a decline in real-world social relationships or mental health. We therefore explored Airman perceptions of social support and resources, both online and offline, using Williams' Internet Social Capital Scales to assess the availability of social support and resources from close friends and family (Williams, 2006).[1] Parallel versions of a ten-item scale measuring perceived access to social support networks asked about (1) people Airmen know exclusively or almost exclusively online and (2) people Airmen know in person ("offline") although they may communicate with them online as well (e.g., a supportive email from a close neighbor would be captured in the offline scale even though it was administered in an online context). One survey question asked respondents to "rate the extent to which you agree or disagree with the following statements. Please note, these questions ask about relationships with people you know *online*. By this, we mean people that you met or only know online, or those that you used to know in person but now

---

[1] We did not use Williams' bridging social capital scales, which assess the breadth of less-personal networks that may open up new avenues of information, in order to reduce the length of the survey and because this issue was less central to the study than bonding social capital.

really only communicate with online." Respondents were asked to rate the following ten statements on a five-point Likert scale, from "strongly disagree" to "strongly agree":

- There are several people online I trust to help solve my problems.
- There is someone online I can turn to for advice about making very important decisions.
- There is no one online that I feel comfortable talking to about intimate personal problems.
- When I feel lonely, there are several people online I can talk to.
- If I needed an emergency loan of $500, I know someone online I can turn to.
- The people I interact with online would put their reputation on the line for me.
- The people I interact with online would be good job references for me.
- The people I interact with online would share their last dollar with me.
- I do not know people online well enough to get them to do anything important.
- The people I interact with online would help me fight an injustice.

These items are designed to function as a scale to measure perceived social capital available online. This scale was repeated, with the word "online" replaced by "offline" (the instructions and the first item clarified that by "offline" we mean "in person"). For each item, a "strongly agree" response received a value of 5, making the highest scale score—reflecting the greatest level of perceived social support—to be 50 (5 x 10 items). As shown in Figure 4.1, more Airmen reported greater social support from their offline (in-person) contacts than from their online contacts (74 percent compared with 12 percent). That difference may not be particularly surprising: what is noteworthy is that more than one in ten Airmen perceive a high level of social support from individuals they know *only* online. This finding speaks to the potential value of ICTs for promoting social well-being among Airmen.

Figure 4.1. Airman Online and Offline Social Support

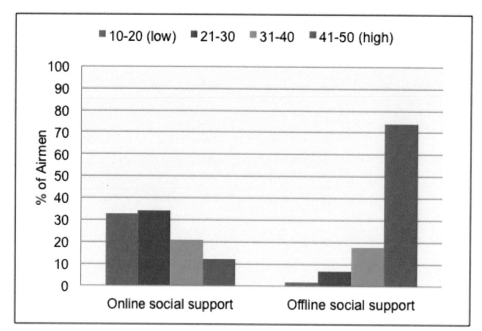

NOTE: N = 3,355 for online support and 3,399 for offline support.

Conclusion

Airmen use a variety of ICTs to make and maintain social ties, including to keep in touch with friends, family, and other Airmen. The types of ICT selected for such connections vary, depending on both the purpose and the recipient. Airmen reported greater social support from their in-person (offline) contacts than from individuals they know only online. This difference may have implications for future intervention efforts, as it suggests that there may be some benefit to fostering localized communities in which individuals may interact face-to-face and social support may be more readily felt. That over ten percent of survey respondents found a high level of social support from people they know only online also suggests that ICTs can play an important role in Airman social well-being.

Chapter Five

The Impact of ICT on Airman Well-Being

Since the mid-1990s, the impact of new ICTs, particularly Internet use, on psychosocial well-being has been much studied, with often-conflicting findings. For example, theories of how Internet use has been thought to affect psychosocial well-being include: (1) it enhances existing relationships, (2) it displaces them, or (3) it differentially impacts people who have fewer social resources or have some existing psychosocial distress, such as loneliness or anxiety (Bessiere et al., 2008).

Part of what underlies discrepant research theories and findings on whether ICT use is beneficial or detrimental may be the sheer number of constructs that relate in some way to subjective well-being. These include, for example, depression, anxiety, loneliness, cognitive distraction, identity, self-esteem, self-efficacy, and creativity. What constitutes use of ICT must also be defined. The effects of ICT use may depend upon the type of activity (e.g., education, gaming, chatting), the amount of ICT use, and the timing of use (e.g., weekday, weekend) (Cotten et al., 2011). This chapter examines the association between ICT use and several measures of well-being, including self-rated mental health, depression, and loneliness, and also examines related constructs of problematic Internet use and cyberbullying, which can have significant implications for mental health and well-being.

Airman ICT Use and Self-Rated Mental Health

Airmen were asked to rate their mental health as excellent, very good, good, fair, or poor. This self-rated item is similar to items used in other research (Fleishman and Zuvekas, 2007; Mawani and Gilmour, 2010). Table 5.1 shows self-rated mental health for the survey respondents. Overall, 83 percent rated their mental health as excellent or very good, and only 5 percent rated their mental health as fair or poor.

Table 5.1. Airman Self-Rated Mental Health

Self-Rated Mental Health	Percent
Excellent	47
Very Good	36
Good	13
Fair	4
Poor	1

NOTE: N = 3,449.

Social Connectedness Through ICT Is Associated with Self-Rated Mental Health

As shown in Table 5.2, the minority of Airmen in our survey who do not use any ICT to keep in touch with friends and family or with other Airmen reported significantly worse mental health than the overall study population ($p < 0.0001$).

Table 5.2. Self-Rated Mental Health of Airmen Who Do Not Use ICT to Keep in Touch with Others

	Excellent or Very Good (percent)	Good (percent)	Fair or Poor (percent)
Do not use any ICT to keep in touch with friends/family (N = 85)	78	12	10
Do not use any ICT to keep in touch with other Airmen (N = 214)	72	15	13
Study population (overall) (N = 3,449)	83	13	4

Relatively Frequent ICT Use Has Little Impact on Self-Rated Mental Health

We also examined self-rated mental health of Airmen who are frequent ICT users relative to other Airmen in our sample.[1] As shown in Table 5.3, there were few differences between the self-rated mental health of frequent ICT users and that of the overall sample.

Table 5.3. Self-Reported Mental Health of Airmen Who Are Frequent ICT Users

ICT Used 20 or More Hours per Week	Self-Rated Mental Health		
	Excellent or Very Good (percent)	Good (percent)	Fair or Poor (percent)
Texting	82	15	4
Email	83	13	4
Phone calls	83	14	3
Social media	82	15	4
Other Internet sites	81	15	4
Video games	80	16	4
IM/chat rooms	82	11	7
Video chat	87	13	0
Blogs	88	12	0
Study population (overall)	83	13	4

NOTE: N ranges from 3,382 to 3,443.

---

[1] About 10 percent of the overall sample for the most commonly used ICTs (see Figure 3.4).

Depression

Research findings on the association between ICT use and depression have been mixed. One explanation for this may lie in the underlying purpose for using the technology. For example, one study found that individuals who use the Internet to communicate with family and friends are less likely to be depressed, but it found that using the Internet to meet new people was associated with increased depression, suggesting that there may be underlying factors related to depression that drive any observed association between Internet use and depression (Bessiere et al., 2008). Interestingly, introverted people with especially few offline resources (e.g., small networks, less social support) who used the Internet to meet people were not among those who experienced increased depression, suggesting that some people may derive particular benefit from using the Internet in certain ways and that the association may be context-specific. Other researchers have stressed that the amount of Internet usage and the timing of usage (e.g., during work hours) should also be taken into account in evaluations of any association with depression or psychological distress (Cotten et al., 2011).

Depressed Mood Among Airmen Was Uncommon and Unrelated to Frequent ICT Use

Figure 5.1 presents Airman responses for each of the two items on the Patient Health Questionnaire 2 (Kroenke, Spitzer, and Williams, 2003) that ask about the extent to which in the past two weeks respondents (1) have "little interest or pleasure in doing things" and (2) are "feeling down, depressed or hopeless." Combining these items into the single measure of depressed mood, we found that 4 percent of the Airmen in the sample were likely to be depressed (i.e., had a score of three or higher out of a possible score of six).

We also found no association between depressed mood and Airman self-reported use of any ICT for 20 or more hours a week.

Figure 5.1. Airman Self-Reported Depressed Mood in the Past Two Weeks

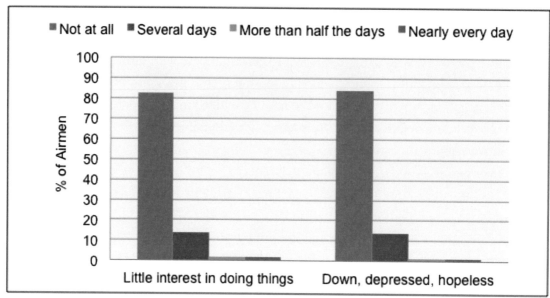

NOTE: N = 3,436.

Specific ICTs Are Associated with Loneliness

The concept of loneliness is important for this study because of concerns about the relationship between ICT use and loneliness and because of the potential impact of loneliness on Airman well-being:

> There is now substantial evidence that loneliness is a core part of a constellation of socioemotional states including self-esteem, mood, anxiety, anger, optimism, fear of negative evaluation, shyness, social skills, social support, dysphoria, and sociability (Hughes et al., 2004 p. 656).

Lonely people may be more likely to use the Internet to improve their mood, for emotional support, or to feel more satisfied with their online interactions than with their in-person interactions (Morahan-Martin and Schumacher, 2003). However, these hoped-for benefits may not be realized and may depend on the purpose for using ICTs. Sheldon (2008) found that Internet use primarily benefited extroverted individuals. Freberg et al. (2010) found that lonely individuals do not actually make more friends online and that online applications (e.g., browsing, forums) that allow anonymous access are most related to increased loneliness. Caplan (2007) argues, however, that social anxiety—and not loneliness—is the confounding variable that is linked with poor well-being outcomes.

To assess loneliness, we used the Three-Item Loneliness Scale (Hughes et al., 2004) derived from the standard 20-item R-UCLA Loneliness Scale used widely for studies in the United States (Shiovitz-Ezra and Ayalon, 2012; Russell, 1996; Russell, Peplau, and Cutrona, 1980). The three items ask

- How often do you feel that you lack companionship?
- How often do you feel left out?
- How often do you feel isolated from others?

Figure 5.2 shows the proportion of Airmen who answered that at the time of the survey, they felt left out or isolated or that they lacked companionship. The score for the loneliness scale is calculated by summing across three items, where "hardly ever" = 1, "some of the time" = 2, and "often" = 3. Airmen who answered "often" to all three questions would have a loneliness scale score of 9. Among the Airmen in our survey, 15 percent scored 6 or higher. This is similar to the proportion (14 percent) in a study of adults ages 50 and over, a population for which loneliness is a socially prevalent phenomenon (Shiovitz-Ezra and Ayalon, 2012).

Figure 5.2. Percentage of Airmen Who Reported Experiencing Loneliness

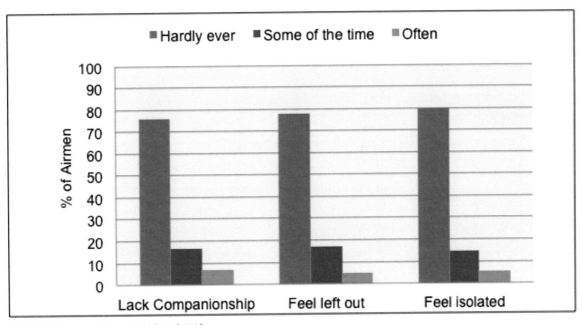

NOTE: N ranges from 3,442 to 3,443.

Individuals who frequently use IM chat rooms or forums, blogs, or social media were slightly more likely to report experiencing loneliness than the overall study population (Table 5.4) ($p < 0.0001$). We cannot assume a causal relationship here; it is possible that rather than the use of these means of communication contributing to feelings of loneliness, those who feel lonely are likely to turn to these sites seeking connection with others or a sense of belonging.

Table 5.4. Percentage of Airmen Who Are Relatively Frequent Users of ICTs Who Experience Loneliness

ICT Used 20 or More Hours Per Week	Experiencing Loneliness (percent)	Not Experiencing Loneliness (percent)
Texting	17	83
Email	13	87
Phone calls	18	82
Social media	21	79
Other Internet sites	18	82
Video games	16	84
IM/chat rooms	23	77
Video chat	19	81
Blogs	22	78
Study population (overall)	15	85

Problematic Internet Use

Although it has been studied for more than 20 years, the question of whether the Internet can be "addictive" is still being debated. Synonyms for or variations of the term "problematic Internet use" include Internet addiction, Internet dependency, excessive Internet use, compulsive Internet use, and pathological Internet use. Researchers and mental health care providers have developed numerous approaches and instruments for measuring and classifying problematic Internet use; one of the earliest studies adapted methods used for the study of pathological gambling (Young, 1996). Like other forms of addiction, Internet use is defined as problematic when it interferes with other life activities and events. Early research on problematic Internet use suggested that its effects include displacing important activities, such as managing one's time (Brenner, 1997). One guide for mental health professionals suggested treatments and preventions for problematic Internet use even while noting the uncertainty of its clinical status (Gray and Gray, 2006). An analysis of clinical data from 929 individuals suggested that problematic Internet use appears to be an outgrowth of preexisting problems, albeit with some Internet-specific differences (Mitchell, Finkelhor, and Becker-Blease, 2007). A meta-analysis of qualitative research reached inconclusive findings as to whether Internet addiction should be considered a clinical disorder (Douglas et al., 2008).

The committee preparing the fifth edition of the *Diagnostic and Statistical Manual of Mental Disorders* (DSM-5) considered adding Internet addiction as a new category (Markel, 2012), but ultimately, the narrower concept of Internet Gaming Disorder was included in an appendix of conditions recommended for further study. The decision to include Internet Gaming Disorder stemmed from the accumulating evidence of similarities to substance-use disorders and gambling disorder documented in the emerging research literature, e.g., tolerance, withdrawal, failed attempts to control use, and impaired functioning (American Psychiatric Association, 2013, pp. 796–798).

Few Airmen Report Problematic Internet Use, But a Small Minority May Be Struggling

To capture the concept of poorly controlled Internet use that has become disruptive in individuals' lives, we used the GPIUS2, the constructs of which have been developed over a number of studies (Caplan, 2002, 2003, 2005, 2007, 2010; Caplan, Williams and Yee, 2009; Davis, 2001). Specific constructs of problematic Internet use measured by this scale are shown in Table 5.5.

Table 5.5. Subscales and Measures in the GPIUS2

Construct	Survey Items
• Preference for online social interaction	• I prefer online social interaction over face-to-face communication. • Online social interaction is more comfortable for me than face-to-face interaction. • I prefer communicating with people online rather than face-to-face.
• Mood regulation	• I have used the Internet to talk with others when I was feeling isolated. • I have used the Internet to make myself feel better when I was down. • I have used the Internet to make myself feel better when I've felt upset.
• Cognitive preoccupation	• When I haven't been online for some time, I become preoccupied with the thought of going online. • I would feel lost if I was unable to go online. • I think obsessively about going online when I am offline.
• Compulsive Internet use	• I have difficulty controlling the amount of time I spend online. • I find it difficult to control my Internet use. • When offline, I have a hard time trying to resist the urge to go online.
• Negative outcomes	• My Internet use has made it difficult for me to manage my life. • I have missed social engagements or activities because of my Internet use. • My Internet use has created problems for me in my life.

SOURCE: Caplan, 2010.

The mean score for problematic Internet use among Airmen was 26 out of a possible 75. This average was similar to the score of 33 found for a nonrandom U.S. sample of 785 people, which is not necessarily a comparable population (Caplan, 2010).[2] Figure 5.3 shows the Airman scores on the GPIUS2. Higher scores indicate greater agreement with the survey items shown in Table 5.5 and thus greater problems with Internet use. The lowest scores indicate little or no Internet use that is poorly-controlled and problematic. While most Airmen fall in the lower ranges of GPIUS2 scores, about 6 percent fall into the top (worst) ranges. If this survey population can be generalized to the total Air Force military population (328,667 as of February 2012), then more than 30,000 Airmen may struggle with problematic Internet use.

Figure 5.3. Airman Problematic Internet Use Scores

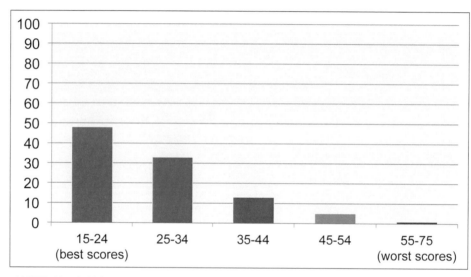

NOTE: N = 3,393.

What this scale does not capture is Airman activities online. Possible compulsive activities include gaming, gambling, shopping, web surfing, online sexual activity, and chatting with others. Additional research would be necessary to identify whether Airmen with problematic Internet use tend to be engaged in any particular type of activity. The scale also does not capture problematic use of other ICTs, such as texting and mobile phone applications.

Airman Problematic Internet Use Is Linked to Well-Being

Airman scores on the GPIUS2 were significantly correlated ($p < 0.0001$) with several measures of their well-being. In particular, as indications of problematic Internet use increased, self-rated mental health decreased ($r = 0.25$), depressed mood increased ($r = 0.18$), and loneliness increased ($r = 0.33$). These associations do not necessarily suggest that problematic

---

[2] Recent prevalence estimates for problematic Internet use among the adult U.S. population are not available.

Internet use leads to worse mental health, however. As noted above, prior research suggests that in some cases, technology addiction may be preceded by preexisting problems, rather than the use of the technology itself causing problems.

Cyberbullying Can Have Negative Mental Health Impacts

The 2014 Pew study found that younger adults are more likely than older adults to experience both kindness and cruelty online (Pew Research Center, 2014). Among Internet users, 44 percent of 18- to 29-year-olds and 21 percent of 30- to 49-year-olds report having been treated unkindly or attacked online (Pew Research Center, 2014, p. 23). Turkle (2011) contends that online anonymity makes it easier to bully, while Internet users' desire to share leads to emotional investment in the feedback, making bullying that much more painful.

Such bullying can have significant negative implications for mental health. Adolescents who have been cyberbullied tend to have low self-esteem (Patchin and Hinduja, 2010). Research conducted on adolescents, although outside the scope of this literature review, indicates that both victims and perpetrators of cyberbullying are more likely to experience suicidal ideation and behaviors than those with no experience with cyberbullying (Hinduja and Patchin, 2010). Cyberbullying may also lead to clues about how pro-self-harm websites, including pro-suicide websites,[3] may be related to suicidal behaviors (boyd,[4] Ryan, and Leavitt, 2010).

About One-Quarter of Airmen Have Had Some Type of Exposure to Cyberbullying in the Past Year

Airmen were asked about their recent experiences with harassment or bullying behavior online, including whether they had witnessed, participated in, or been victims of such behavior. These questions—and the subsequent results—did not distinguish between Airman experiences with cyberbullying in general and cyberbullying specifically within the Air Force or related to other Airmen. Given ongoing concerns about harassment within the military, future research may focus specifically on this topic.

As shown in Table 5.6, 25 percent of Airmen reported some type of experience with cyberbullying in the past year. Relatively fewer Airmen reported having intervened in cyberbullying (13 percent) than having witnessed such behavior (23 percent). Even fewer Airmen reported either being the victim of (7 percent) or participating in (3 percent) cyberbullying behaviors.

---

[3] On pro-suicide websites, individuals encourage one another to kill themselves, provide information about the methods for doing so, and even make suicide pacts (Biddle et al., 2008).

[4] Note that boyd's legal name is spelled with all lower-case letters.

Table 5.6. Airman Experience with Cyberbullying

Experience	Percent
I have witnessed someone being harassed, bullied, "flamed," or otherwise attacked online	23
I have been harassed, bullied, "flamed," or otherwise attacked online	7
I have participated in harassing, bullying, "flaming," or otherwise attacking someone online	3
I have intervened when I witnessed someone else being harassed, bullied, "flamed," or otherwise attacked online	13
I have not had any experience with this behavior in the past 12 months	75

NOTE: N = 3,408.

Further analyses of these data show that Airmen who have witnessed cyberbullying were less likely to rate their own mental health as excellent (37 percent) than those who have not witnessed cyberbullying (50 percent) ($n = 3,406$)($p < 0.0001$).

Conclusion

For the majority of Airmen, the regular, and even frequent, use of ICT is not associated with poorer mental health or depressive symptoms. In fact, Airmen in our survey who did not use any ICT to keep in touch with friends and family or with other Airmen reported significantly worse mental health than that of the overall study population. Although frequency of ICT use was not associated with our measures of mental health or depressive symptoms, the results do suggest that individuals engaging in greater ICT use experience a higher level of loneliness than those who use ICT less frequently. Self-reports in our survey suggest that ICT use is, for the most part, not harmful for mental health and well-being and may in fact be beneficial in moderation and if used to connect with friends and family.

However, the survey did identify two areas of particular concern that may warrant additional exploration. First, a small minority of Airmen reported a problematic level of Internet use. Furthermore, as problematic Internet use increased, self-rated mental health decreased, depressed mood increased, and loneliness increased, suggesting that interventions specific to individuals with problematic Internet use may be worth further consideration. The survey also identified concerns around cyberbullying, as one-quarter of Airmen had either witnessed, experienced, or perpetrated such activity within the 12 months prior to the survey. Although almost half of them reported intervening when they had witnessed cyberbullying, efforts to stop such behavior and encourage intervention when it is observed may be beneficial, particularly as the uptake and applications of ICTs continue to spread.

Chapter Six

Airmen's Use of ICT to Seek and Receive Mental Health Information and Services

A primary interest of the Air Force is to understand whether ICT should be better leveraged to promote mental health and reduce barriers, and if so, how that should be accomplished. ICT is used not only for entertainment or to socialize with others but also to seek information. This chapter examines ways in which Airmen have used ICT to seek and receive mental health information and services (telehealth or telemental health) and would be willing to use it for that purpose in the future. It reports how Airmen seek mental health information, the types of sources they indicate using, and how they use the information they obtain. It then discusses Airman self-reported skills in locating, evaluating, and applying that information. The chapter next focuses on Airmen as recipients of mental health information sent by others through ICT. Finally, it addresses Airman preferred communication modes for Air Force outreach, discussing mental health concerns with others, and receiving advice, care, or treatment from a mental health professional.

Fewer Airmen Seek Mental Health Information Than Physical Health Information

Websites, applications for mobile phones and tablets, and virtual worlds provide a wide range of health information and resources. Pew found that in 2012, 72 percent of 18- to 29-year-olds and 67 percent of 30- to 49-year-olds said they looked online for health information (Fox and Duggan, 2013, p. 14).

Our study findings for seeking information related to *physical* health are similar to the Pew findings: 80 percent of Airmen reported using ICT to find such information, primarily on Internet sites (72 percent, see Table 6.1). Airmen also reported using email (14 percent), social media (13 percent), and phone calls (11 percent) to obtain information on physical health. Roughly 7 percent reported that they had used chat rooms, blogs, and texting. Although we had anticipated that our ICT category of "video games, including virtual worlds and social games such as Small World, Second Life, World of Warcraft" would be an unlikely source of health information for most, if not all, respondents, we included it on the survey to capture a potential baseline for future studies. Although none of our respondents reported using video games or virtual worlds to obtain health information, these resources are beginning to appear on the virtual world Second Life. As of February 2013, Second Life included an education center from the American Cancer Society; a Mayo Clinic virtual conference facility and bookstore; HealthInfo

Island, which presents information on health and wellness issues; and centers associated with universities (Second Life, undated).

Many fewer Airmen reported using ICT to access *mental health* information (Table 6.1). More than half (54 percent) reported that they had not used any ICT to find information about mental health. Almost all of those who did use an ICT for this purpose used the Internet (42 percent).

As with physical health, we included the video game/virtual world option as a possible source of mental health information, although we did not anticipate high levels of use. One effort to provide mental health information in Second Life was developed by the Department of Defense (DoD) National Center for Telehealth and Technology (T2). The T2 Virtual PTSD Experience invites participants to "endure the causes, confront the symptoms, and discover the help available for combat-related PTSD in this serious role-play adventure" (Second Life, undated). None of our respondents reported using this type of ICT to find mental health information.

Table 6.1. Airman Use of ICT to Find Physical and Mental Health Information

ICT	Physical Health Information (percent)	Mental Health Information (percent)
Texting	6	3
Email	14	8
Phone calls	11	7
Social media	13	6
Other Internet sites	72	42
Video games	0	0
IM/chat rooms	7	5
Video chat	1	1
Blogs	7	4
Did not do this	20	54

NOTE: N ranges from 3,443 to 3,447.

When the question was narrowed to activities in just the past 12 months, 68 percent—more than two-thirds—indicated they had not used any ICTs to learn about mental health topics or resources, including suicide prevention. About one-quarter (24 percent) had used the Internet, 12 percent had used email, 7 percent reported using the phone, and 7 percent had used social media to learn about mental health. However, some respondents may have forgotten that they learned about mental health through ICTs several months prior to the survey, and it is also possible that some Airmen who reported seeking mental health information actually conducted their searches more than 12 months ago. Thus, these results should be read as a general indication, rather than an exact accounting of Airman behavior in the 12 months prior to the survey.

We followed up with the Airmen who said they had used ICT to learn about mental health in the past 12 months and asked if they sought information about suicide prevention specifically.

Nearly one-third (31 percent) of the respondents, 355 Airmen, reported that they had used ICT to learn specifically about suicide prevention. Of these, 77 percent indicated that Air Force suicide prevention training was at least part of this information. Six percent of our overall survey sample reported learning about suicide prevention through ICT, separate from required Air Force training.

Airmen Use Diverse Sources to Find Information on Mental Health and Suicide Prevention

We asked the Airmen who reported searching for information on mental health and suicide prevention to list the sources they go to most often. Of those who reported seeking out at least one source for mental health information (23 percent of all respondents, N = 796), 15 percent listed Air Force websites, 8 percent listed Air Force mental health professionals, 8 percent cited Air Force training or briefings, and 3 percent noted chain-of-command messaging (e.g., commander's calls). One-fifth (22 percent) named DoD's Military OneSource as a resource, and 4 percent listed Department of Veterans Affairs (VA) resources. The most popular civilian resources listed were WebMD (26 percent), search engines such as Google and Yahoo (25 percent), and websites associated with health insurance companies, government agencies (e.g., the National Institute of Mental Health), and nonprofit organizations (e.g., the Mayo Clinic) (15 percent). Additionally, 5 percent of respondents named websites associated with commercial talk shows or health-focused magazines. Information-aggregation sites, such as Wikipedia and Reddit, were mentioned by 6 percent of those who listed a mental health resource.

Overall, responses to questions about sources of information on suicide prevention were similar, with a few exceptions. Of the Airmen who described seeking out at least one source for information about suicide (3 percent of the total sample, N = 93), 18 percent listed Air Force websites, 8 percent listed Air Force mental health professionals, 3 percent cited Air Force training or briefings, and 4 percent noted chain-of-command messaging (e.g., commander's calls). One-fifth (19 percent) named Military OneSource as a resource, and 11 percent listed VA sources. The most popular civilian resources cited were search engines such as Google and Yahoo (17 percent) and websites associated with health insurance companies, government agencies (such as the National Institute of Mental Health), and nonprofit organizations (e.g., the Mayo Clinic) (13 percent); information-aggregation sites, such as Wikipedia and Reddit were cited by 5 percent. Only 2 percent mentioned WebMD or websites associated with commercial talk shows or health-focused magazines. Thus, Airmen are more likely to name VA resources and less likely to name WebMD and search engines when seeking information on suicide prevention, as compared with seeking information on mental health more generally. They are using military-affiliated or -supported resources but are not limiting their searches to those resources.

Airmen Use the Mental Health Information They Seek Through ICT

Some of the Airmen who reported using ICT to find information on mental health and suicide prevention used that information to promote their own mental health or the mental health of their friends or family. Table 6.2 shows that about 42 percent of Airmen shared the information they found. Beyond sharing the information, many reported using it to make critical decisions specific to their mental health care—seeking help (20 percent), managing mental health concerns (18 percent), and considering treatment options, both on their own (15 percent) and in conjunction with a mental health professional (11 percent).

In addition to using information found via ICT, a minority reported leveraging ICT to share thoughts of sadness or loneliness (8 percent) or suicide (3 to 4 percent) with friends or health professionals. Other ways of using the information reported through write-in comments on the surveys included for general knowledge, for work-related purposes, and to help others who were having problems.

Table 6.2. How Airmen Use Mental Health Information They Seek Through ICTs

Use of Information	Percentage of Respondents
Share with friends or family	42
Change the way I manage my mental health concerns	18
Decide whether I should see a mental health professional	20
To make a decision about the best treatment option for me	15
Share thoughts of sadness or loneliness with a friend or peer	12
Talk with a mental health professional about the information	11
Share thoughts of sadness or loneliness with a health/mental health professional	8
Share my thoughts about suicide with a friend or peer	4
Share my thoughts about suicide with a health/mental health professional	3
Other	39

NOTE: N = 1,030.

Airmen's Self-Rated eHealth Literacy Skills Differ

High-quality, clearly written health information from reputable sources can be valuable for individuals looking to learn more about a condition, determine if they should seek professional care, or inform decisions about preferred courses of treatment. Yet it can be difficult to distinguish high- from low-quality information. Websites can have erroneous, out-of-date, or unclear information, which may be misleading or perhaps may even steer readers to harmful decisions. The complexity of the website and its layout, along with the use of technical terms and medical jargon, can also render websites less useful for informational purposes, even if the quality of the information itself is high (Eysenbach et al., 2002).

We sought to assess Airmen's self-reported eHealth literacy, that is, how well they believe they can navigate the electronic health environment. For the survey, we modified items from a general eHealth literacy scale to focus on mental health specifically (Norman and Skinner, 2006b). Airmen were asked to report the degree to which they agreed or disagreed (using a five-point Likert scale) with the following statements:

- I know how to find helpful mental health resources on the Internet.
- I know how to use the mental health information I find on the Internet to help me.
- I have the skills I need to evaluate the mental health resources I find on the Internet.
- I can tell high-quality from low-quality mental health resources on the Internet.
- I feel confident in using information from the Internet to make mental health decisions.

Possible scores on this scale ranged from 5 (strongly disagree with all of the above statements) to 25 (strongly agree with all of the above statements). As shown in Figure 6.1, only 45 percent of Airmen are on the most desirable end of the spectrum. The scores for the remainder reflect some uncertainty about ability to find, evaluate, and use mental health information and resources available online.

Figure 6.1. Airman Self-Reported eHealth Literacy for Mental Health Resources

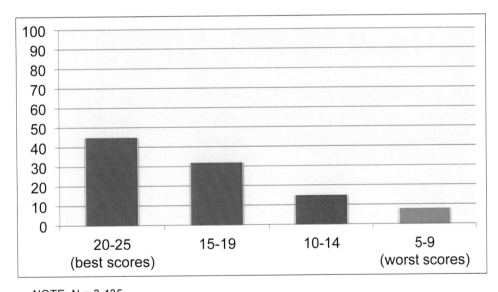

NOTE: N = 3,435.

More work is needed to understand the relation between eHealth literacy skills, the use of ICT to find information, and how that information is used to promote mental health. For example, the extent to which Airmen are leveraging the eHealth literacy skills of their family and social networks or advising others in their networks is not clear. Also, the scale used in our survey assesses Airmen's *perceptions* about their abilities, not whether they actually *can* distinguish high-quality from low-quality information.

Airmen Receive and Use Mental Health Information Others Send Them Through ICT

The survey findings provide significant insight into the active seeking of health information and how Airmen understand and use such information. Active seeking can be conceptualized as a "pull" approach—information is available online and on base for Airmen to find and use. A complementary strategy can be conceptualized as a "push" approach—the Air Force sending information directly to Airmen.

We asked Airmen about mental health information they have received in the past 12 months and what they did with that information. Since this question asked them to recall communication from the past year, the responses may underreport these experiences, as Airmen may have forgotten information they received some time ago. Asked, "In the past 12 months, has anyone used any technologies to send you information about mental health or mental health care?" 30 percent replied that they had (N = 1,142). The vast majority of them (91 percent, N = 390) reported that they had read, watched, or listened to the information. Additionally, 72 percent reported finding the information helpful, 8 percent reported that it was not helpful, and 20 percent did not know (N = 389). Collectively, 32 percent of Airmen reported that the information prompted them into action (N = 388).

To examine the efficacy of "pushing" information out to Airmen, we asked survey participants "which ways would be most helpful for receiving mental health outreach, educational, or screening materials from the Air Force." Respondents could endorse as many modes of communication as they wished. For this item and for the remaining items discussed in this chapter, we added "face-to-face" as an option. Also, we dropped "video games" from the "combined video games and virtual worlds" option, so that it read, "online virtual worlds like Second Life, Small Worlds." Two-thirds of the Airmen reported that they would like to receive such mental health information from the Air Force via face-to-face interaction, but 56 percent also reported that receiving such information via email would be helpful (see Table 6.3). Some Airmen felt that receiving such information via Internet sites (36 percent) and social media (24 percent) would be most helpful. Although blogs, texting, and IM/chat rooms were not the most popular options, they were chosen by approximately one in ten Airmen. Other suggestions entered by respondents included regular mail, pamphlets or flyers, classes, and briefings.

Table 6.3. Preferences for Receiving Mental Health Information from the Air Force

Communication Mode	Percentage of Respondents
Face-to-face	67
Email	56
Other Internet sites	36
Social media	24
Phone calls	16
Blogs	11
Texting	11
IM/chat rooms	10
Online virtual worlds	5
Video chat	5

NOTE: N = 3,408.

The results were similar when we narrowed our analyses to only the respondents who rated their own mental health fair or poor or who indicated depressed mood (approximately 200 Airmen in our sample).

Online Connections with Others Provide Additional Sources of Health Information and Support

In addition to contacting mental health professionals or experts in the field, individuals sometimes turn to one another for advice and support. A 2013 nationwide survey indicated that in the past year, 24 percent of adults received support or information online from others who have the same health condition as they do, and 26 percent learned about someone else's health issues online (Fox and Duggan, 2013, pp. 3, 5). Moreover, 16 percent of Internet users went online specifically to look for others who have the same health concerns (Fox and Duggan, 2013, p. 5). Beyond the exchange of information, such connections may also help to provide a sense of community and foster social connections, social support, and social capital, which are critical to well-being.

Online support groups provide a forum where people can participate anonymously, without indicators of status (such as rank, wealth, race), without worrying about how they look or sound, and with the freedom to leave and later reenter the discussion if feelings become intense and they need time to process (Barak, Boniel-Nissim, and Suler, 2008). The benefits of such groups to overall well-being can include the psychological impact of writing, the expression of emotions, the sharing of information, the development of relationships, and influence on decisionmaking and action, such as making medical appointments or seeking a second opinion on a diagnosis (Barak, Boniel-Nissim, and Suler, 2008). Support groups are not a substitute for treatment, and they may have drawbacks as well, including inappropriate behavior by members, the spread of misinformation, and reinforcement of maladaptive belief systems (Barak, Boniel-Nissim, and Suler, 2008).

41

We asked Airmen to report on their preferences for discussing mental health concerns with others who have similar concerns. As noted in Chapter Four, 35 percent of the survey sample had used ICT to connect with others who have similar health concerns. The vast majority of all Airmen in our sample (75 percent) reported that they would find face-to-face interactions helpful (see Table 6.4). About one-quarter to one-third felt that email (34 percent), social media (26 percent), phone calls (25 percent), and chat rooms (24 percent) would also be helpful for speaking with others who have similar concerns. The preferences for Airmen with low self-rated mental health or depressed mood are very similar to the overall preferences reported in Table 6.4.

Table 6.4. ICT Preferences for Discussing Mental Health Concerns with Others Who Have Similar Concerns

Communication Mode	Percentage of Respondents
Face-to-face	75
Email	34
Social media	26
Phone calls	25
IM/chat rooms	24
Other Internet sites	21
Blogs	17
Texting	12
Video chat	11
Online virtual worlds	6

NOTE: N = 3,394.

Airmen Are Interested in ICTs to Supplement Face-to-Face Interactions with Mental Health Professionals

ICTs have great potential to help overcome many psychological and logistical barriers to seeking mental health care. They can help break down barriers to care such as concern about stigma and privacy (being seen walking into a counselor's office) and logistical challenges, such as inconvenient office locations and hours, particularly for those living in remote areas. Although more research is needed, some psychological interventions delivered over the Internet have yielded positive treatment outcomes (Andersson and Culjpers, 2009; Barak et al., 2008) and are becoming increasingly common with the expansion of telehealth and telemedicine. Research on the delivery of exposure therapy for PTSD through telehealth shows that it can actually reduce symptoms not only of PTSD but also of stress, anxiety, depression, and other impairments (Gros et al., 2011; Hilty et al., 2013). Despite the potential of ICTs to supplement and expand service delivery, caution is warranted, as not all individuals are ideal candidates for technology-based intervention (e.g., high-risk individuals who may pose harm to themselves or others), and the ICTs themselves pose logistical challenges and problems of provider licensing across state lines.

Appendix E presents an overview of the guidelines for telemental health care practice developed by a number of mental health care professional associations.

Though the use of ICTs to deliver care is relatively nascent, it is gaining momentum. When we asked Airmen about ways that would be most helpful for seeking *advice* from a mental health care professional for themselves or others, 83 percent reported that meeting face-to-face would be most helpful (Table 6.5). Again, Airmen were able to select as many communication options as they deemed "most helpful." Emails and phone calls (38 percent each) and other Internet sites (22 percent) for such consultations were endorsed at nowhere near the rate that face-to-face interactions were endorsed. Still, as of February 2012, 10 percent of the Air Force military population amounted to 32,867 Airmen. If those Airmen share ICT preferences at rates seen in Table 6.2, then even texting, video chat (Skype, VTC), and blogs offer opportunities to extend professional advice to many in the Air Force community.

Table 6.5. Communication Modes Considered Most Helpful for Seeking Advice from a Mental Health Care Professional

Communication Mode	Percentage of Respondents
Face-to-face	83
Email	38
Phone calls	38
Other Internet sites	22
IM/chat rooms	15
Social media	13
Texting	10
Video chat	9
Blogs	9
Online virtual worlds	4

NOTE: N = 3,422.

Airmen expressed less interest in receiving mental health *care or treatment* via ICTs, however. More than 90 percent reported that face-to-face was the most helpful way to receive treatment (Table 6.6). Only 20 percent endorsed the second most commonly selected means of interaction, phone calls, which was followed by email (16 percent). Even fewer (9 percent or less) felt that other options would be most helpful.

Table 6.6. Communication Modes Considered Most Helpful for Receiving Mental Health Care or Treatment

Communication Mode	Percentage of Respondents
Face-to-face	91
Phone calls	20
Email	16
Video chat	9
Other Internet sites	9
IM/chat rooms	7
Social media	6
Texting	5
Blogs	4
Online virtual worlds	2

NOTE: N = 3,430.

Although many Airmen selected more than one mode of communication, about two-thirds (63 percent) checked *only* "face-to-face" as most helpful for mental health care or treatment. This finding, coupled with the other risks and challenges mentioned above, suggests that ICTs should best be conceptualized as supplemental to, not in place of, face-to-face provider interaction for mental health concerns.

Additionally, almost one in ten of our respondents (9 percent) *did not* check "face-to-face" as most helpful for mental health care or treatment but checked one or more other options. If that response is representative of the Air Force population, nearly 43,000 Airmen believe that ICT would be preferable to in-person communication for mental health care or treatment. Table 6.7 presents the communication modes preferred by this subset of the sample. We do not know whether these respondents are thinking of a hypothetical situation or feel they have an existing need (or whether that matters) or what type of circumstances they may be referencing. Perhaps their answers already account for or would differ for a range of needs, such as guidance while coping with an adverse life event like a divorce or the death of a loved one, routine management of psychiatric prescriptions, treatment for PTSD, counseling to improve self-esteem or communication skills, or counseling to help heal following a sexual assault.

Table 6.7. ICTs Considered Most Helpful for Receiving Mental Health Care or Treatment by Airmen Who Do Not Believe Face-to-Face Interaction Would Be Most Helpful

ICT	Percentage of Respondents
Email	47
Phone calls	31
Other Internet sites	24
Social media	19
Texting	17
IM/chat rooms	10
Blogs	8
Video chat	7
Online virtual worlds	4

Again, we explored whether the ICT preferences for mental health advice (reported in Table 6.5) and for mental health care or treatment (reported in Table 6.6) are different for Airmen who reported fair or poor mental health or were likely depressed, but we found that the results are remarkably similar.

Conclusion

Despite a wealth of high-quality mental health information available on the Internet and via various other ICTs, more than half of the Airmen in our survey reported not using any ICTs to seek out such information, compared with only 20 percent who said they did not use ICTs to find physical-health information. When asked to focus on only the past 12 months, approximately one-third of the Airmen indicated that they had used ICTs to learn about mental health topics or resources, including suicide prevention. Moreover, 11 percent of the respondents said they learned about suicide prevention through ICT, separate from required Air Force training. Given the potential for recall bias, those numbers could reflect an underestimation of this activity.

Airmen who did seek mental health information in the past year, however, reported a number of positive behaviors as a result, including the sharing of information with others who may be at risk, making changes to the way in which they are managing their own mental health concerns, and deciding to seek treatment. The low proportion of Airmen reporting the use of ICTs to find information and resources about mental health may be driven, in part, by challenges related to their ability to find information, distinguish high- from low-quality information, and use the information to make health-related decisions or to inform health behaviors. However, it may also reflect a lack of need for this type of information. Previous research has found that time spent online seeking health information has been associated with increased depression, possibly because of greater need or greater rumination or anxiety about potential health problems (Bessiere et al., 2010; Cotten et al., 2011).

In addition to asking about current information-seeking practices, the survey also addressed Airman willingness to use ICTs for future receipt of mental health information and for mental

45

health treatment. Although respondents indicated that they are willing to use ICT for connecting with others with similar concerns, speaking with providers to obtain advice, and receiving mental health care and treatment, most still prefer face-to-face interaction, suggesting that ICT should supplement, but not replace, in-person interaction. The ICT mental health communication preferences of Airmen with fair or poor self-rated mental health and/or likely depression were similar to those of the overall survey sample. Given that telemedicine is a relatively new concept, these opinions may change over time as ICT use for these purposes becomes more common. This may be an area of future interest for the Air Force, particularly to enhance the support of reserve and guard members who may be located further from Air Force or DoD resources.

Chapter Seven

Airmen's Perceptions of the Relationship Between ICT and Their Social and Mental Well-Being

This chapter provides an overview of the results of three survey items reflecting Airmen's own views on the importance of various ICTs for their social and mental well-being.

Our sample included many Airmen with recent assignments away from home:

- 21 percent of them had been deployed overseas 30 days or more in the past year.
- 4 percent were deployed at the time of the survey.
- 26 percent had been on a TDY assignment or had another temporary geographic relocation away from their home station (excluding deployments) for more than 30 days in the past year.

Previous research with military personnel has found that communication with friends and family back home can provide social support and boost morale, although it can be disruptive or distressing when the deployed personnel confront problems or when expected communication is lacking (for a review, see Greene et al., 2010). Our survey asked all participants, "Which technologies are most important for your social and mental well-being when you are TDY, deployed, or have another temporary geographic relocation away from your home station?" As shown in Table 7.1, more than 80 percent of respondents reported phone calls and email as important for their social and mental well-being while away from home. Video chat was selected by 64 percent of Airmen—a much higher endorsement than video chat received for other survey items. The second highest endorsement of video chat was presented in Table 4.1, which showed that 37 percent of Airmen use it to keep in touch with friends and family.

Among the written comments submitted with this survey item were remarks that cards, letters, and packages were also important for respondents' well-being when they were away from home.

**Table 7.1. ICTs Most Important for Social and Mental Well-Being
When Away from Home**

ICT	Percentage
Phone calls	89
Email	82
Video chat	64
Social media	62
Texting	55
Other Internet sites	25
IM/chat rooms	20
Blogs	5
Video games	5
Online virtual worlds	5

NOTE: N = 3,431.

The survey then asked only those who were deployed at the time of the survey (N = 124) or had deployed more than 30 days in the past year (N = 685) to rate the impact of the ICTs on their social and mental well-being during their most recent deployment. While ICTs can connect service members with loved ones during deployments, they can also bring home-front problems to the battlefront. Thus Airmen were asked to rate the ICTs on a five-point Likert scale from "mostly positive" to "mostly negative" but were also given options to indicate that an ICT was not available during deployment (which, at 30 percent, was the case for text messaging more than any other technology) or that it was available but they did not use it. As shown in Figure 7.1, almost none of our respondents perceived a negative impact. Indeed, ICTs were reported as generally available and beneficial during deployment.

Figure 7.1. Perceived Impact of ICTs on Social and Mental Well-Being During Airmen's Most Recent Deployment

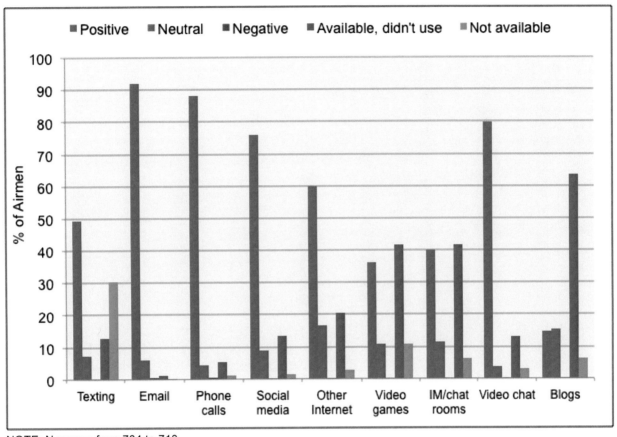

NOTE: N ranges from 704 to 713.

The final survey question asked all participating Airmen to reflect on the impact of these ICTs on their overall well-being and life satisfaction today. On a five-point Likert scale, from "mostly positive" to "mostly negative," most Airmen perceive either a positive or a neutral impact (Figure 7.2). The ICTs Airmen most often rated as negative were video games (6 percent), blogs and social media (5 percent each), and IM (4 percent).

Figure 7.2. Perceived Impact of ICTs on Airman Overall Well-Being and Life Satisfaction

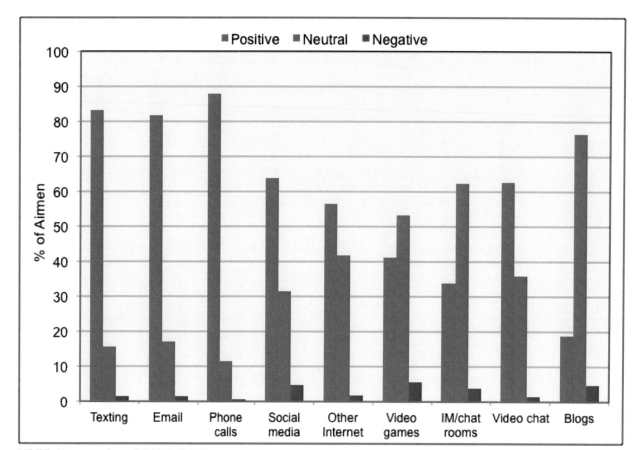

NOTE: N ranges from 3,368 to 3,435.

Conclusion

Most Airmen perceive that many ICTs are important to their social and mental well-being, both when they are deployed or assigned away from home and in general. The perceived impact of the ICTs is predominantly positive. Phone calls and email top the lists, and for those who are away from home, video chat and social media follow in importance. Overall, Airmen view text messaging as positive as well, but it is not as available in the deployed environment.

Chapter Eight
Demographic Differences in ICT Use and Perceived Well-Being

We conducted additional analyses to examine demographic differences for our key variables, because of their potential implications for Air Force targeted outreach efforts, both for whom to target and for what ICTs might be effective channels for communication.

The demographic-differences analyses consisted of multivariate logistic regressions, using the weighted sample described in Chapter Two, with demographic and military characteristics included as predictors. By including multiple variables, we were able to examine the effects of one demographic characteristic while accounting for the effects of the others. For some of these characteristics, we had to collapse some of the survey-response categories because of the low number of Airmen in particular subgroups. To ensure sufficient sample size for our comparisons, we collapsed across subgroups when a subgroup had fewer than 200 respondents. Specifically, the regression model included the following variables:

- Component: active, guard, reserve
- Rank: enlisted, officer
- Gender: male, female
- Age: 18–24, 25–34, 35–44, 45+
- Relationship status: never married, in a relationship with a boyfriend or girlfriend, married, divorced/separated/widowed
- Time away from home: a binary category of (1) those deployed at the time of the survey and/or deployed overseas or on TDY for more than 30 days in the previous year and (2) those who were not
- Children: children under the age of 18, no children under 18
- Education level: high school diploma or GED, associate's degree or vocational/technical diploma, bachelor's degree, graduate degree
- Race/ethnicity: non-Hispanic white, minority race/ethnicity.

Although we ran these analyses for most of the variables, scales, and response options presented in this report, we were unable to do so for some of them. For instance, we did not run analyses for questions with open-ended responses, for follow-up questions asked of a subset of respondents that were of insufficient size to support this type of analysis, and for the response options of "other" or "other Internet sites." Specifically, we examined demographic differences for the following variables: frequent ICT use; online and offline social capital; self-rated mental health; depressed mood; loneliness; problematic Internet use; experiences with cyberbullying; mental health information-specific ICT use; eHealth literacy; attitudes toward preferred communication modes for Air Force outreach; discussing mental health concerns with others;

receiving advice, care, or treatment from a mental health professional; which ICTs are most helpful when away from home; and the impact of ICTs on well-being and life satisfaction. A subset of these variables had dichotomous response options, while others had ordinal response options. For the dichotomous variables, we conducted binary logistic regressions; for the ordinal variables (i.e., problematic Internet use, loneliness, depressed mood, self-rated mental health, offline and online social capital, and impact of ICTs), we conducted ordinal logistic regressions.

In this chapter, we report only statistically significant differences ($p < 0.01$) that may also have some practical difference for the Air Force.[1] We chose this p value, which is conservative by most standards, to help compensate for the alpha inflation caused by the large number of differences tested. Demographic differences were not significant unless otherwise noted.

The tables in this chapter present the key findings. Each table indicates which groups were more likely to endorse particular survey items, for dichotomous variables, and more likely to endorse higher-value items, for ordinal variables. (See Appendix A for the complete wording of survey items and response options.)

Each table also provides the multivariable-adjusted odds ratios and p values, for scientific audiences that are interested in greater detail. Odds are equal to the probability that an event occurs relative to the probability that it does not occur, i.e., $P/(1 - P)$. An odds ratio is the ratio of the odds between two groups, i.e., a head-to-head comparison of odds. When the event probability is rare, the odds ratio is approximately equal to the relative probabilities of the event of interest between groups. As an example, Table 8.2 (later in this chapter) shows that the odds ratio for a negative mental health rating between enlisted Airmen and officers is 1.38, which means that enlisted Airmen have 38-percent greater odds of a negative mental health self-rating than officers. If negative mental health self-ratings are infrequent, we could also say that enlisted Airmen have a 38 percent *greater* likelihood of providing a negative mental health rating than enlisted officers.

In the following sections, we describe demographic differences by component, rank group, gender, age group, relationship status, time away from home (TDY or deployed), and whether Airmen have minor children. There were no consistent statistically and practically significant differences for the demographic variable of majority/minority race and ethnicity. That is to say, Airmen generally had similar ICT use and social and psychological well-being patterns regardless of their level of education and whether they were a racial or ethnic minority or majority.

---

[1] An example of a statistically significant difference that may not be practically significant is the difference in email use for work purposes: 82 percent of active-duty Airmen indicated that they use email every day for work purposes, while 74 percent of guard and 78 percent of reserve Airmen indicated the same. There are no apparent practical implications of those differences for Air Force policy or practice.

Component Differences

Table 8.1 presents our findings for component differences, in terms of which ICTs were used, preferred, and believed to be important or beneficial. Active-duty Airmen were more likely than guard and reserve Airmen to report frequent use of video games (20 or more hours a week) and were more likely than guard to report video games as having a positive impact on their overall well-being and life satisfaction. Active-duty Airmen were also more likely than guard and reserve Airmen to report higher levels of online social capital (social support from others they interact with online), whereas guard members were more likely to report higher levels of offline social capital (social support from others they interact with in person). Active-duty Airmen were more likely than reserve to report social media as among the most important ICTs when they are on TDY, deployed, or have another temporary geographic relocation away from their home stations. Active-duty Airmen were also more likely than guardsmen to report video chat, such as Skype, as among the most important ICTs when traveling away from home and more likely than both guard and reserve to report video chat as having a positive impact on their overall well-being and life satisfaction.

Guard member's preferences for ICT communication related to mental health differed from those of active-duty Airmen in a number of ways. Guardsmen were more likely to indicate that email, texting, and IM/chat rooms/forums would be most helpful for receiving mental health outreach, educational, or screening materials from the Air Force. Guard members were also more likely to report phone calls and texting as most helpful for receiving mental health care or treatment and were also more likely to report email as most helpful for discussing mental health concerns with others and to report phone calls, email, and texting as most helpful for seeking advice from a mental health care professional for themselves or others.

Although there were no component differences in ICTs used to learn about mental health topics, reserve Airmen were more likely than active-duty Airmen to share the mental health information they obtained via ICTs with family and friends. Reserve Airmen, compared with active-duty Airmen, were like the guard in that they were also more likely to report IM/chat rooms/forums as most helpful for receiving mental health outreach from the Air Force. Both guard and reserve Airmen were more likely than active-duty Airmen to value text messaging when on TDY, deployed, or otherwise away from their home stations.

Thus, for the most part, component differences tell us most about how best to reach Airmen to support their mental health, as their responses to psychological well-being items are similar once other demographic differences are taken into account.

Table 8.1. Component Differences in ICT Use, Perceived ICT Value, and Well-Being Measures

Subgroup	Survey Item	Odds Ratio	p value
Active duty	Frequent user of video games		
	– compared with guard	2.44	0.001
	– compared with reserve	2.08	0.003
	Video chat has positive impact on well-being/life		
	– compared with guard	1.69	<0.001
	– compared with reserve	1.35	0.004
	Video games have positive impact on well-being/life (compared with guard)	1.61	<0.001
	Video chat is important when deployed or away (compared with guard)	1.59	<0.001
	Social media are important when deployed or away (compared with reserve)	1.37	0.005
	Online social capital		
	– compared with reserve	1.33	0.002
	– compared with guard	1.28	0.005
Guard (compared with active duty)	Texting helpful for Air Force mental health outreach	1.72	0.001
	Texting helpful for seeking advice from mental health professional	1.71	0.001
	Texting helpful for receiving mental health care	1.71	0.007
	IM helpful for Air Force mental health outreach	1.60	0.009
	Texting important when deployed or away	1.59	<0.001
	Phone calls helpful for receiving mental health care	1.55	<0.001
	Email helpful for Air Force mental health outreach	1.45	<0.001
	Phone calls helpful for seeking advice from mental health professional	1.38	0.002
	Email helpful for seeking advice from mental health professional	1.34	0.005
	Email helpful for discussing mental health concerns	1.34	0.006
	Offline social capital	1.30	0.003
Reserve (compared with active duty)	Share ICT mental health information with family/friends	1.90	0.002
	Texting important when deployed or away	1.84	<0.001
	IM helpful for Air Force mental health outreach	1.79	0.002

Officer and Enlisted Differences

Table 8.2 presents the results of our comparisons of the responses of officers with those of enlisted personnel. Enlisted Airmen differed from officers in ways that suggest possible greater need for mental health support and greater preferences for certain ICTs to provide that support. Enlisted Airmen were more likely than officers to report worse mental health. They were also more likely to report greater online social capital, but they reported similar offline (in-person) social capital. Enlisted Airmen were more likely than officers to report using video chat and social media in the past 12 months to learn about mental health topics or resources, including suicide prevention. Most notably, enlisted Airmen were much more likely to indicate that the mental health information they found through ICTs led them to share their thoughts about suicide with health or mental health professionals. Enlisted Airmen were more likely than officers to report texting as one of the most helpful ICTs for both receiving mental health outreach, educational, or screening materials from the Air Force and receiving mental health care or treatment. Furthermore, enlisted Airmen were more likely than officers to report face-to-face as one of the most helpful means of communication for discussing mental health concerns with others who have similar concerns. Finally, enlisted Airmen were more likely to feel that IM/chat rooms/forums are among the most important ICTs when on TDY, deployed, or away from their home stations.

Table 8.2. Enlisted Differences from Officers in ICT Use, Perceived ICT Value, and Well-Being Measures

Subgroup	Survey Item	Odds Ratio	p value
	ICT mental health info led to sharing thoughts about suicide with professionals	18.22	0.003
	Learn about mental health via video chat	6.65	0.005
	Learn about mental health via social media	2.56	0.005
	Texting helpful for receiving mental health care	2.49	0.002
Enlisted	Texting helpful for Air Force mental health outreach	1.90	0.006
	IM important when deployed or away	1.86	<0.001
	Face-to-face helpful for discussing mental health concerns	1.48	0.008
	Online social capital	1.45	0.001
	Negative self-rated mental health	1.38	0.008

Gender Differences

Table 8.3 presents our findings for gender differences. Women were more likely than men to report worse mental health, yet they also reported higher levels of both online and offline social capital. Women were more likely to be frequent users of email, texting, and social media and to believe that those ICTs and phone calls have a positive impact on their overall well-being and life satisfaction. Women were also more likely than men to report social media as among the most important ICTs when deployed or otherwise away from their home stations.

Men were more likely than women to be frequent users of video games and to believe that video games have a positive impact on their overall well-being and life satisfaction. Men were more likely than women to report witnessing and intervening in cyberbullying, while women were more likely to report having had no experience with cyberbullying.

Women were more likely than men to report social media as most helpful for discussing mental health concerns with others who have similar concerns and to report face-to-face as one of the most helpful modes for receiving mental health care or treatment. Men were more likely than women to prefer phone calls and video chat for receiving mental health outreach, educational, and screening materials from the Air Force.

Table 8.3. Gender Differences in ICT Use, Perceived ICT Value, and Well-Being Measures

Subgroup	Survey Item	Odds Ratio	p value
Women	Face-to-face helpful for receiving mental health care	2.00	0.001
	Frequent user of texting	1.85	<0.001
	Frequent user of social media	1.66	0.002
	Frequent user of email	1.66	0.002
	Texting has a positive impact on well-being/life	1.65	<0.001
	No experience with cyberbullying	1.63	<0.001
	Believe phone calls have a positive impact on well-being/life	1.56	<0.001
	Believe email has a positive impact on well-being/life	1.56	<0.001
	Negative self-rated mental health	1.55	<0.001
	Social media important when deployed or away	1.53	<0.001
	Social media has positive impact on well-being/life	1.52	<0.001
	Social media helpful for discussing mental health concerns	1.50	<0.001
	Offline social capital	1.43	<0.001
	Online social capital	1.31	0.003
Men	Frequent user of video games	3.70	<0.001
	Video games have a positive impact on well-being/life	2.33	<0.001
	Video chat helpful for Air Force mental health outreach	2.08	0.006
	Witness cyberbullying	1.61	0.001
	Intervene in cyberbullying	1.59	0.009
	Phone calls helpful for Air Force mental health outreach	1.59	0.003

Age-Group Differences

Table 8.4 presents our findings for differences among age groups. The most numerous differences among subgroups of Airmen were found between those ages 18 to 24 and those in other age groups. Airmen 18 to 24 years old were more likely than older Airmen to report signs of problematic Internet use; to perceive greater offline social capital; and to report witnessing, experiencing, participating, and intervening in cyberbullying. There were no significant age-group differences for our measures of loneliness, depressed mood, or self-rated mental health.

We found that Airmen ages 18 to 24 were more likely than older Airmen to be frequent users (20 or more hours a week) of texting, video games, and social media (hours spent on each ICT are not necessarily mutually exclusive). Younger Airmen were more likely than older Airmen to report using blogs to learn about mental health topics and resources and to believe that blogs have a positive impact on their well-being and life satisfaction. Younger Airmen were also more likely to report texting, social media, IM/chat rooms/forums, and blogs as most helpful for receiving mental health outreach from the Air Force. Similarly, younger Airmen were more likely than older Airmen to report social media, IM/chat rooms/forums, and blogs as most helpful for discussing mental health concerns with others who have similar concerns.

There were also some age-group differences regarding the value of certain ICTs when traveling or deployed and for overall well-being and life satisfaction. Younger Airmen were more likely than older Airmen to report texting, social media, and IM as most important for their social and mental well-being when TDY, deployed, or otherwise temporarily relocated away from their home stations, while Airmen in each of the age groups over 24 were more likely than younger Airmen to report email as most important in those situations. Airmen ages 18 to 24 were also more likely than older Airmen to believe that each of the ICTs with the exceptions of phone calls and email has a positive impact on their overall well-being and life satisfaction.

Table 8.4. Age-Group Differences in ICT Use, Perceived ICT Value, and Well-Being Measures

Subgroup	Survey Item	Odds Ratio	p value
	Intervene in cyberbullying, compared with		
	– 25- to 34-year-olds	16.67	0.002
	– 35- to 44-year-olds	1.79	0.003
	– 45+	1.85	0.008
	Participate in cyberbullying, compared with		
	– 35- to 44-year-olds	5.26	<0.001
	– 45+	7.69	0.001
	Experience cyberbullying (compared with 45+)	4.35	<0.001
	Learn about mental health via blogs (compared with 45+)	3.23	0.009
	Blogs have a positive impact on well-being/life (compared with 45+)	2.94	<0.001
	Frequent user of texting, compared with		
	– 35- to 44-year-olds	2.70	<0.001
	– 45+	3.85	<0.001
	IM helpful for discussing mental health concerns (compared with 45+)	2.63	<0.001
	IM helpful for Air Force mental health outreach (compared with 45+)	2.63	0.001
	Social media has positive impact on well-being/life (compared with 45+)	2.56	<0.001
	Blogs helpful for Air Force mental health outreach (compared with 45+)	2.44	<0.001
	Frequent user of video games, compared with		
	– 35- to 44-year-olds	2.44	0.003
	– 45+	7.14	<0.001
	Social media important when deployed or away, compared with		
	– 35- to 44-year-olds	1.64	0.003
	– 45+	3.03	<0.001
18- to 24-year-olds	Video chat important when deployed or away, compared with		
	– 25- to 34-year-olds	2.08	<0.001
	– 35- to 44-year-olds	3.23	<0.001
	– 45+	5.00	<0.001
	Video chat has a positive impact on well-being/life, compared with		
	– 35- to 44-year-olds	2.38	<0.001
	– 45+	3.57	<0.001
	Texting important when deployed or away, compared with		
	– 25- to 34-year-olds	1.92	<0.001
	– 35- to 44-year-olds	2.22	<0.001
	– 45+	2.56	<0.001
	Frequent user of social media, compared with		
	– 35- to 44-year-olds	2.22	0.001
	– 45+	6.25	<0.001
	Social media helpful for discussing mental health concerns (compared with 45+)	2.22	<0.001
	Blogs helpful for discussing mental health concerns, compared with		
	– 35- to 44-year-olds	1.92	0.002
	-- 45 year olds	2.50	<0.001
	Video games have a positive impact on well-being/life, compared with		
	– 25- to 34-year-olds	1.59	0.002
	– 35- to 44-year-olds	2.08	<0.001
	– 45+	4.35	<0.001
	Texting helpful for Air Force mental health outreach, compared with		
	– 25- to 34-year-olds	2.00	0.003
	– 35- to 44-year-olds	2.04	0.005
	-- 45+	2.33	0.001
	IM has a positive impact on well-being/life (compared with 45+)	2.00	<0.001

Subgroup	Survey Item	Odds Ratio	p value
	Social networking sites helpful for Air Force mental health outreach (compared with 45+)	1.89	0.002
	Texting has positive impact on well-being/life, compared with		
	– 35- to 44-year-olds	1.67	0.001
	– 45+	2.27	<0.001
	Problematic Internet use, compared with		
	– 35- to 44-year-olds	1.61	0.008
	– 45+	2.56	<0.001
	Witness cyberbullying, compared with		
	– 25- to 34-year-olds	1.54	0.003
	– 35- to 44-year-olds	1.56	0.007
	– 45+	2.17	<0.001
	Offline social capital (compared with 45+)	1.54	0.003
25- to 34-year-olds	Email helpful when deployed or away (compared with 18- to 24-year-olds)	2.20	<0.001
35- to 44-year-olds	Email helpful when deployed or away (compared with 18- to 24-year-olds)	3.68	<0.001
45+	Email helpful when deployed or away (compared with 18- to 24-year-olds)	4.76	<0.001

Differences by Relationship Status

Table 8.5 presents our findings for differences by relationship status. Airmen who have never married and Airmen who are divorced, separated, or widowed were more likely to report high levels of loneliness, depressed mood, and worse mental health than married Airmen. There were no consistent, significant differences between married Airmen and Airmen in a relationship with a boyfriend or girlfriend. In addition, Airmen who have never married were more likely to report problematic Internet use and to be among the most frequent users of social media. Divorced, widowed, or separated Airmen (combined into a single subgroup) were more likely than married Airmen to be frequent users of texting and social media. Moreover, divorced, widowed, or separated Airmen were more likely than married Airmen to report using phone calls to learn about mental health topics or resources, to believe that virtual worlds would be most helpful for receiving outreach from the Air Force, and to feel that that texting is one of the most important ICTs for well-being when deployed or temporarily located away from home. Married Airmen, on the other hand, were more likely than divorced, widowed, or separated Airmen to be frequent users of blogs and more likely than Airmen who have never married to report video chat as most important when deployed or away. Married Airmen were also more likely than never-married Airmen to respond that email and texting have a positive impact on their well-being and life satisfaction and to report greater offline social capital. Never-married Airmen were more likely than married Airmen to report video chat as having a positive impact on their overall well-being, and divorced, widowed, or separated Airmen were more likely to report texting as having a positive impact.

58

Table 8.5. Differences by Relationship Status in ICT Use, Perceived ICT Value, and Well-Being Measures

Subgroup	Survey Item	Odds Ratio	p value
Never married (compared with married)	Loneliness	4.34	<0.001
	Frequent user of social media	1.96	0.004
	Depressed mood	1.64	0.003
	Video chat has a positive impact on well-being/life	1.57	0.002
	Negative self-rated mental health	1.46	0.009
	Problematic Internet use	1.47	0.004
Divorced, separated, or widowed (compared with married)	Loneliness	3.44	<0.001
	Depressed mood	1.79	0.001
	Negative self-rated mental health	1.74	0.001
	Frequent user of texting	2.49	<0.001
	Virtual worlds helpful for Air Force mental health outreach	2.35	0.005
	Learn about mental health via phone calls	2.18	0.003
	Frequent user of social media	1.93	0.009
	Texting important when deployed or away	1.54	0.009
	Texting has a positive impact on well-being/life	1.53	0.006
Married	Email has a positive impact on well-being/life (compared with never-married)	1.79	<0.001
	Offline social capital (compared with never-married)	1.59	0.001
	Texting has a positive impact on well-being/life (compared with never-married)	1.52	0.004
	Video chat important when deployed or away (compared with never-married)	2.44	<0.001
	Frequent user of blogs (compared with divorced, separated, or widowed)	111.11*	<0.001

* No divorced, separated, or widowed Airmen reported being frequent blog users.

Differences by Time Away From Home

Table 8.6 presents our findings for differences by time away from home. Airmen who were deployed at the time of the survey or who had spent more than 30 days away from home in the past year for deployments or temporary duties were more likely to report using phone calls to learn about mental health topics or resources, including suicide prevention. Compared with Airmen who had not spent that amount of time away from home in the past year, these Airmen were also more likely to report video chat and social media as among the most important ICTs for their social and mental well-being when on TDY, deployed, or on another temporary geographic relocation away from their home stations. Airmen who had spent time away were also more likely to believe that social media, video games, IM, and blogs have a positive impact on their well-being. Airmen who have not spent time away from home, on the other hand, were more likely to report virtual worlds as being helpful for discussing mental health concerns with others with similar concerns and to report video chat as having a positive impact on well-being.

Table 8.6. Differences by Time Away from Home in ICT Use, Perceived ICT Value, and Well-Being Measures

Subgroup	Survey Item	Odds Ratio	p value
Deployed at the time of the survey and/or deployed overseas or on TDY for more than 30 days in previous year	Learn about mental health via phone calls	1.58	0.008
	Blogs have a positive impact on well-being/life	1.42	0.001
	Video chat important when deployed or away	1.41	<0.001
	Social media has a positive impact on well-being/life	1.40	<0.001
	Video games have a positive impact on well-being/life	1.29	0.003
	Social media important when deployed or away	1.28	0.006
	IM has positive impact on well-being/life	1.49	<0.001
Not deployed at the time of the survey and/or deployed overseas or on TDY for more than 30 days in previous year	Virtual worlds are helpful for discussing mental health concerns	1.79	0.007
	Video chat has a positive impact on well-being/life	1.54	<0.001

Differences by Having or Not Having Children

Table 8.7 presents our findings for differences by having or not having children. Airmen with children under the age of 18 were more likely to report video chat as one of the most important ICTs for their social and mental well-being when on TDY, deployed, or on another temporary geographic relocation away from their home stations. Airmen without minor children were more likely to report being frequent users of video games and more likely to use virtual worlds to learn about mental health topics or resources.

Table 8.7. Differences by Having or Not Having Children in ICT Use, Perceived ICT Value, and Well-Being Measures

Subgroup	Survey Item	Odds Ratio	p value
Children under age of 18	Video chat important when deployed or away	1.49	<0.001
No children under age of 18	Learn about mental health via virtual worlds	11.21	0.005
	Frequent user of video games	2.06	0.002

Education-Level Differences

Table 8.8 presents our findings for education-level differences. The few education-level differences that we found arose primarily around blogs and video chat being more likely to be preferred by Airmen with higher education levels. Airmen with a graduate degree were more likely than Airmen with a high school degree or GED to report using video chat to learn about mental health topics or resources and to believe that blogs would be helpful for receiving mental health outreach from the Air Force. Airmen with a high school diploma or GED, however, were more likely than those with a graduate degree to report blogs as having a positive impact on their well-being and life satisfaction. Moreover, Airmen with a graduate degree and those with a Bachelor's degree were more likely than those with a high school diploma or GED to report blogs as being among the most helpful ICTs for discussing mental health concerns with others who have similar concerns. Both Airmen with an Associate's degree and those with a Bachelor's degree were more likely to report video chat as being most helpful for seeking advice from a mental health professional than those with a high school diploma or GED. Airmen with an Associate's degree were also more likely than those with a high school diploma or GED to report video chat as being helpful for receiving mental health care or treatment. Finally, both Airmen with an Associate's degree and those with a graduate degree were more likely than those with a high school diploma or GED to report blogs as being important to their social and mental well-being when they were deployed or away from home.

Table 8.8. Education-Level Differences in ICT Use, Perceived ICT Value, and Well-Being Measures

Subgroup	Survey Item	Odds Ratio	p value
Graduate degree (compared with high school/GED)	Learn about mental health via video chat	9.38	0.002
	Blogs important when deployed or away from home	2.89	0.009
	Blogs helpful for receiving mental health outreach from Air Force	2.42	<0.001
	Blogs helpful for discussing mental health concerns	2.20	0.001
Bachelor's degree (compared with high school/GED)	Video chat helpful for seeking advice from mental health professional	2.07	0.005
	Blogs helpful for discussing mental health concerns	1.64	0.008
Associates degree (compared with high school/GED)	Blogs important when deployed or away from home	2.22	0.009
	Video chat helpful for seeking advice from mental health professional	1.85	0.006
	Video chat helpful for receiving mental health care	1.84	0.004
	Blogs have a positive impact on well-being/life	2.08	<0.001
High school degree (compared with graduate degree)	Blogs have a positive impact on well-being/life	2.08	<0.001

Conclusion

Although Airman overall responses on this survey were very similar, there were some statistically significant differences between demographic subgroups that held even when we accounted for the effects of other differences. Men, active-duty Airmen, and Airmen ages 18 to 24 were more likely to be frequent video game users (20 or more hours a week, or in the top 10 percent of Airmen in terms of frequency). Frequent texting was more likely among women, Airmen ages 18 to 24, and divorced, widowed, or separated Airmen. Those same subgroups, along with never-married Airmen, were more likely to be frequent users of social media.

We also found some differences in responses on social and psychological well-being items. Active-duty, enlisted personnel, and women reported greater online social capital relative to their comparison groups, whereas guard members, women, and 18- to 24-year-olds were more likely to report greater social capital among those they interact with in person. Enlisted Airmen, women, and never-married, divorced, separated, and widowed Airmen were more likely to self-rate their mental health negatively. Men and Airmen ages 18 to 24 were more likely to have had experiences with cyberbullying. Problematic Internet use, or use that may be poorly controlled, compulsive, and causing problems in their lives, was more likely to be indicated by Airmen ages 18 to 24 and Airmen who have never married. Additionally, never-married Airmen and divorced, widowed, or separated Airmen were more likely to report loneliness and depressed mood. Our findings that younger Airmen were particularly likely to use the newer ICTs and to have experiences with cyberbullying are consistent with the findings of previous research done in the general population (Lenhart et al., 2008; Pew, 2014). The findings are also consistent with earlier literature on demographic differences in mental health that find that enlisted (Lapierre, Schwegler, and LaBauve, 2007; Martin, 2007; Smith et al., 2008; Tanelian and Jaycox, 2008), female (Hoge, Auchterlonie, and Milliken, 2006; Smith et al., 2008) and non-married (Lapierre, Schwegler, and LaBauve, 2007; Smith et al., 2008) members of the military were particularly likely to suffer from negative mental health.

The analyses reported in this chapter document different ICT preferences that can guide tailoring of ICT for specific mental health efforts by the Air Force. For example, email, texting, and phone calls are more likely to be welcome avenues for mental health-related interactions among guard populations than among active-duty Airman. Airmen ages 18 to 24 are more likely than older Airmen to be interested in receiving mental health outreach, educational, and screening materials from the Air Force through blogs, text messaging, social media, and IM, chat rooms, or forums. Enlisted Airmen were more likely than officers to select text messaging as a preferred ICT for mental health outreach and for mental health care or treatment.

Chapter Nine

Leveraging Airman Use of ICT to Promote Social and Psychological Well-Being

This chapter provides highlights of the results of our 2012 survey, followed by a discussion of how ICT may offer new avenues for identifying and reaching out to Airmen in distress and for promoting social and psychological health across the Air Force. The chapter also describes the limitations of this study and points to particularly fruitful directions for additional inquiry. Finally, it summarizes a number of recommendations the Air Force may wish to consider as it works to further support the psychological health of Airmen.

Overview of Survey Results

The survey queried Airmen about their use of various ICTs: texting, email, phone calls, social media, other Internet sites, video games, IM and chat rooms, video chat, and blogs. The responses from the 3,479 participating Airmen were weighted to reflect the Air Force military population according to affiliation (active, guard, reserve), gender, rank group (officer, enlisted), and age group (18–24, 25–34, 35–44, 45 and older). The sample size was sufficient for the analyses presented in this report, but because the response rates were low, particularly among Airmen in the youngest age group, we recommend that future efforts experiment with different survey invitation and administration modes and assess whether variation has any impact on response rates and on the survey responses themselves.

Not surprisingly, we found that personal use of ICT is common among Airmen. Overall, 29 percent of the respondents reported using one or more ICTs for more than 20 hours a week for nonwork purposes. Some subgroups of Airmen are more likely to be frequent users of specific ICTs—for example, men, active-duty Airmen, and Airmen ages 18 to 24 were more likely than other respondents to be frequent video game users. Many Airmen use ICTs to keep in touch with friends, family, and other Airmen, and most perceive a positive or neutral impact of such use on their well-being. Twelve percent perceived a high level of social support from their online contacts, and this was more likely to be reported by active-duty Airmen, enlisted Airmen, and Airmen who are women. ICTs are perceived to be particularly valuable during deployment: For example, 80 percent of Airmen reported that video chat had a positive impact on their social and mental well-being during their most recent deployment. Almost no Airmen self-reported a negative impact of these ICTs during deployment.

Although problematic Internet use is uncommon among Airmen, a small minority (6 percent) may be struggling—thinking obsessively about going online, having difficulty controlling use, turning to the Internet when feeling down, and finding that the Internet has created problems in their lives. The survey revealed that as problematic Internet use increases, self-rated mental health decreases, and signs of depressed mood and loneliness increase. Airmen ages 18 to 24 and never-married Airmen were more likely than their counterparts to score negatively on the problematic Internet use scale.

About one-quarter of Airmen have had some exposure to cyberbullying in the past year. Males and Airmen ages 18 to 24 were more likely to have these experiences than other Airmen. Most of those who had such exposure were witnesses to others being bullied, but 7 percent were themselves targets. Thirteen percent of Airmen intervened when they witnessed cyberbullying, and 3 percent admitted to participating in cyberbullying. Airmen who had witnessed cyberbullying were less likely to rate their own mental health as excellent than those who had not witnessed cyberbullying.

Airmen use a variety of ICTs to seek mental health information, with the Internet being the ICT used most. Six percent of the Airmen surveyed reported using ICT to learn about suicide prevention separately from Air Force training in the past 12 months. Airmen passed along mental health information acquired through ICT and used it to make decisions about mental health care. However, only about half of the survey respondents believe they can evaluate online mental health resources and have confidence in using information from the Internet to make mental health decisions.

Face-to-face interaction is by far the preferred mode for mental health outreach, discussing mental health concerns with others who have similar concerns, and receiving mental health care advice and treatment. But a sizable minority prefer the option of communicating through ICT. Moreover, nearly 10 percent indicated that they would prefer a communication mode other than face-to-face for mental health care or treatment. Some demographic differences emerged in this area: For example, enlisted Airmen were more likely than officers to select text messaging as a preferred ICT for mental health care or treatment.

ICTs Offer New Avenues to Identify Suicide Risk and Provide Mental Health Support

Initial research suggests that ICTs may provide opportunities to identify and intervene with individuals who are particularly at risk for suicide. Those who seek online support or information are higher-risk (e.g., more depressive symptoms, less in-person social support) than those who do not go online for such information (Harris, McLean, and Sheffield, 2009). A number of innovations have been developed to leverage social networks and ICT to reach individuals who are at high risk or in crisis for suicide. We describe several of these below, not to suggest that the Air Force replicate them, but as examples of the potential of ICTs to accomplish goals related to a reduction in suicides. The Air Force could also coordinate with the VA, which has been developing evidence-based telemedicine treatments for PTSD and other mental health problems.

The It Gets Better Project is an online organization that aims to prevent suicides of people who have been bullied for being gay (itgetsbetter.org). Its website allows people to upload supportive videos, and it has attracted videos from celebrities, professional athletes, businesses, organizations, and even President Barack Obama. Another resource, the Trevor Project, is a free and confidential crisis and suicide prevention service that provides several channels to support sexual minorities, including a crisis phone line; a secure online messaging service for those who are suicidal; an online, non-urgent question-and-answer service; and a social networking community (thetrevorproject.org). Google and Facebook also flag potential at-risk behaviors on their services and attempt to connect users to suicide prevention services. (A list of some online suicide resources is provided in Appendix F.)

Scholars at the Center for the Study and Prevention of Suicide at the University of Rochester have developed an automated data-collection program to map the online social networks of lesbian, gay, and bisexual youth ages 16 to 24, a population that tends to be isolated and at higher risk for suicide. The maps can support peer-driven intervention for harder-to-reach populations and help curtail harmful attitudes and behaviors, such as suicide contagion (Silenzio et al., 2009). Additionally, SAHAR is an online suicide prevention service designed to provide immediate, anonymous support to severely distressed, suicidal users. According to one assessment, SAHAR has provided direct feedback to police that saved more than 4,000 suicidal people, without a single suicide by anyone who had been referred to it (Barak, 2007).

Based on early indications, such suicide prevention approaches through the Internet are promising, but they need further evaluation. Referring Airmen to some of the more trusted medical websites and working with general search-engine companies, such as Google, to increase the prominence of Air Force or DoD resources in search results may be worth further consideration. One particular challenge, however, is the dearth of quantitative research on the effectiveness of many suicide prevention websites and resources. Though a limited number of studies find positive effects, the study methods and outcomes of interest vary considerably (Stone, Barger, and Potter, 2005; Miller and Gergen, 1998; Greidanus and Everall, 2010; Harris, McLean, and Sheffield, 2009; Jong, 2004; Haas et al., 2008). One review of suicide prevention approaches using technologies such as telephone, text messaging, videoconferencing, and Internet concluded that anecdotal and qualitative evidence provides initial support for the view that ICT-based approaches to suicide prevention can be effective, but many unresolved issues still exist, including cultural appropriateness and lack of controlled trials to establish causation (Krysinska and De Leo, 2007). More research would be needed to assess the effectiveness of many of the suicide prevention websites prior to affiliation or endorsement by the Air Force.

Several Airmen wrote suggestions for ways they would like to make anonymous inquiries to a credible professional for mental health outreach, discussions, care, or treatment. These suggestions included a call-in line, IM, and an online forum. Such options, especially if anonymous, could be a way for the Air Force to connect with Airmen who would like mental health assistance but might not otherwise seek it, such as those who fear being seen walking into an appointment, those who are unsure whether mental health professionals could help with their particular problems, and those who are unsure whether their counseling needs could jeopardize their security clearances. Additional research could clarify how Airmen might use an anonymous, interactive communication channel to discuss their concerns with mental health professionals before their problems reach the crisis level. An Air Force–sponsored resource would be able to address questions about Air Force and non–Air Force options for care.

ICTs Offer New Avenues to Promote Well-Being More Broadly

Air Force leaders face a tension between workplace policies and practices that permit Airmen to spend too much time using ICT for personal use (e.g., texting friends or visiting Facebook instead of focusing on the mission) and those that cut off ways for Airmen to sustain their social support networks and quality of family life while meeting their duty requirements. To help Air Force leaders strike that balance, the Air Force should educate leaders and supervisors about the potential benefits of responsible personal ICT use during the work day and should provide them guidance on acceptable local policies regarding the use of ICT. Leaders should also understand that for some Airmen, an obsessive use of ICT could signal the need for intervention to address underlying behavioral or emotional issues, and thus disciplinary action may not be the best approach for resolving problematic levels of use.

The Air Force might also consider ways in which new ICTs could be used to enhance strategic communication approaches to increase awareness of mental health resources, encourage responsible help-seeking, and decrease the stigma of mental health care. Only about half of the Airmen in our survey reported having the skills to identify, evaluate, and confidently use online mental health information and resources, suggesting two avenues for potential intervention: (1) help Airmen develop eHealth literacy skills so that they can find relevant and high-quality information on their own, and (2) ensure that high-quality information and materials are easy to find, access, understand, and use. More work is needed to understand the relation between eHealth literacy skills, the use of ICT to find information, and how that information is used to promote mental health. For example, the extent to which Airmen are leveraging the eHealth literacy skills of their family and social network or advising others in that network is not clear.

At a minimum, the Air Force might consider ways in which mental health information could be delivered that is easy to find and understand and that comes from credible and trusted sources. This effort should include a review of existing websites and resources to make sure information is consistent and consolidated (e.g., that the websites do not take users to multiple sources for different pieces of information) and that it includes current contact information for resources so that users are not frustrated by dead ends, such as broken links, missing web pages, and phone numbers or other resources that are no longer in service. Due to the nature of the Internet and the information age, finding high-quality, clearly written health and medical information can be difficult for anyone. It can be particularly challenging for those struggling with depressed mood or suicidal thoughts, and any effort to ensure that the best information is the easiest to find and access is going to be in the best interest of Airmen. Existing resources the Air Force could leverage include an online guide sponsored by the National Institutes of Health (nlm.nih.gov/medlineplus/healthywebsurfing.html) as well as a free, online 16-minute tutorial (nlm.nih.gov/medlineplus/webeval/webeval.html). Also, the nonprofit Spry Foundation has developed a detailed and clearly written guide for evaluating online health information (spry.org/sprys_work/education/EvaluatingHealthInfo.html).

Additional Research Could Inform Air Force Implementation of These Recommendations

This study only began to explore Airman ICT use, social support, and well-being and potential Air Force use of ICT to bolster its mental-care system. By necessity, this project was bounded to include only the highest-priority research questions, although many other important questions remain. Future research could build on the current study findings and help the Air Force further develop its services to meet the needs of Airmen and their families.

Additional research could also help to identify how usage rates and/or effects on social support and well-being cluster (i.e., "hang together"). Schrock (2008) proposes technology clusters as a means of understanding whether the likelihood of use of an ICT is based on similar perceived characteristics of another ICT or medium. These clusters may include, for example, ICTs for one-to-one communication (e.g., IM, text messaging, email), many-to-many communication (e.g., social media, photo sharing, discussion boards), downloads of content (e.g., movies or music), types of devices (e.g., computer, mobile phone), or information-seeking services (e.g., search engines, maps). Different demographic subgroups or different types of challenges (e.g., PTSD, complicated grief, child behavioral issues, marital conflict) may also be linked with variations in Airman patterns and preferences for seeking and receiving support.

Further research is also necessary to identify the types of activities that are consuming the time of a small minority of Airmen who are engaged in a problematic type of Internet use. The social and psychological well-being of these Airmen, as well as their ability to be productive members of the Air Force, may be in jeopardy if this behavior is not addressed. Qualitative research would permit an in-depth exploration of why some Airmen prefer online social interaction over face-to-face interaction; why they turn to the Internet when they are feeling isolated, depressed or upset; what kind of thoughts about being online preoccupy them; what strategies they have already tried to control their Internet use; and what, if anything, they have done to try to mitigate the negative outcomes caused by their Internet use. Additional in-depth study could also address with a greater level of sophistication how Airmen with mental health concerns—either for themselves or for their family members—believe the Air Force might be able to further develop telemental health resources to support them. Further research is also necessary to understand the use of ICT by Airmen who are in distress in order to improve detection and intervention and to potentially prevent suicides.

Conclusion

- As in the general population, Airmen typically use ICT to augment interactions with real-world friends, family, and other Airmen, not to replace them.
- Six percent of Airmen may be facing serious difficulties due to their poorly controlled and disruptive Internet use. Problematic Internet use was significantly correlated with self-rated poor mental health, depressed mood, and loneliness.
- Airmen seek and use online mental health information, but they need help evaluating its quality.
- Most Airmen prefer face-to-face communication with health professionals, but a sizable minority may prefer ICT alternatives.

Recommendations

The results of this exploratory survey suggest that the Air Force could help Airmen by the following actions:

- Promoting policies that permit nondisruptive on-duty access to ICT to help strengthen family ties, social support networks, and well-being
- Enhancing strategic communication plans to increase Airman awareness of mental health resources, encourage responsible help-seeking, and decrease the stigma of mental health care
- Continuing to support access to ICT during deployment
- Informing leaders that ICT "addiction" may be a sign of broader problems and encouraging appropriate referrals
- Educating Airmen about how to recognize signs of problematic ICT use and the resources that are available to help them address it
- Teaching Airmen, particularly young Airmen, how to protect themselves against and respond to cyberbullying
- Providing basic guidelines for identifying credible websites and directing Airmen to credible sources through messaging and Air Force web pages
- Developing/leveraging alternative communication modes such as texting, social media, and Internet sites to supplement, but not replace, face-to-face interaction between Airmen and mental health care professionals
- Integrating into Air Force suicide prevention training information about the potential role of ICT for identifying suicide risk factors (e.g., via messages indicating intentions for self-harm) and Airmen's options for responding to suicidal messages, especially when those who receive them do not personally know the individuals at risk or where they are located
- Exploring targeted mental health outreach through ICT during deployment
- Exploring through further research the experiences, needs, and attitudes of Airmen experiencing problematic ICT use

- Exploring through further research how Airmen in distress use ICT, in order to improve the Air Force's ability to detect and intervene when Airmen are at risk and thus potentially save lives
- Exploring through further research how Airmen would like to use ICT for mental health support, for what types of problems, and why
- Exploring ways to utilize ICTs to enhance mental health treatment adherence and symptom reduction.

Appendix A

Survey Instrument

This appendix contains the online survey instrument, along with the computer programming notes that indicate items that should be displayed based upon previous responses.

Technologies to Promote Social and Psychological Well-Being in the Air Force

Instructions on screen:

The Air Force is interested in how technologies such as the Internet, mobile phones, and social media can affect the social and mental well-being of Airmen. The Air Force Surgeon General, Lt Gen Charles B. Green, has commissioned the RAND Corporation, a non-profit research company, to conduct a survey to reveal Airmen's attitudes and activities related to this topic. You have been randomly selected to represent Airmen in this study. In this survey, we will ask you some questions about what technologies you use, what you use them for, and what you think about them. We will also ask you a few questions about your own social interactions and mental health. Finally, we will ask you about using technologies to access information and support services to enhance the social and mental fitness of yourself and others. This survey takes about 15-20 minutes to complete.

All of your answers are completely confidential and your individual responses will never be shared with the Air Force or the DoD. No one in the Air Force will know if you have participated in this study or not. No Air Force leaders or health professionals will have access to your individual responses, and none will receive reports about the responses from their units. You do not have to take this survey. No negative consequences to your assignments, promotions, or benefits will result if you choose not to participate. If you do choose to volunteer for this survey but find there are any questions you do not want to answer, please feel free to skip them. You may also stop taking the survey at any time. Your candid responses to as many questions as possible, however, will help ensure that Air Force leadership correctly understands positive and negative aspects of these technologies and takes effective approaches to related policy and support service outreach.

If you are unable to complete the survey in a single session, you may return before the survey closes on [Insert date here]. To protect your answers, however, three days after you start the survey we will seal your responses and allow access only to the unfinished portion of the survey. If you have any technical issues in taking this survey, please contact the survey helpdesk at ntwebhelp@rand.org or 877-260-9246.

If you have any questions about the purpose or content of the survey, please contact Dr. Laura Miller, lauram@rand.org, 703-413-1100 extension 5912. If you have any questions or concerns about your rights as a research subject, please contact Dr. Tora Bikson, tora@rand.org, 310-393-0411 extension 7227.

If you or anyone you know is in emotional crisis, confidential help is available through the Military Crisis Line: call 1-800-273-8255 and press 1, or send a text to 838255, or visit the website to initiate a confidential chat http://veteranscrisisline.net/ActiveDuty.aspx]

Please indicate your consent to participate in this study:

1 I have read the above statement about this study and volunteer to participate.
2 I do not want to participate in this study → [move to exit message]

Instructions on screen: Thank you for agreeing to participate. Before we begin it would be helpful to know a little more about you and your affiliation with the Air Force.

MODULE 1A: INITIAL DEMOGRAPHICS *(additional demographics collected at end of survey)*

DEM01. Which is your Air Force affiliation?
1 Active Duty Air Force (active component)
2 Air National Guard
3 Air Force Reserve
4 Other → **STOP survey.** *Instructions on screen:* Thank you for your time. However, this survey is intended for individuals affiliated with the Active Duty Air Force, Air National Guard, or Air Force Reserve. *EXIT TO HOME SCREEN*

DEM02. What is your current paygrade?
1 E1-E4
2 E5-E6
3 E7-E9
4 O1-O3
5 O4-O6
6 O7-O10

DEM03. What is your gender?
1 Male
2 Female

DEM04. What is your age group? *[Note: these age groups are consistent with the Air Force Personnel Center's Interactive Demographic Analysis System*
1 18-24
2 25-34
3 35-44
4 45+

DEM05. NOT including any deployments, have you been TDY or had another temporary geographic relocation away from your home station for more than 30 days in the past year? **Select one.**

1 Yes
2 No

DEM06. Are you currently deployed? **Select one.**

1 Yes
2 No

DEM07. Have you been deployed overseas for more than 30 days in the past year? **Select one.**

1 Yes
2 No

DEM08. Which one of these operations was your most recent deployment in support of? *[Programmer note: Show only to those who said yes to DEM06 and/or DEM07]*

1   Operation Iraqi Freedom/Operation New Dawn
2   Operation Enduring Freedom
3   Other

Instructions on screen: Thank you. We would now like to ask you about technologies that you use. Please think about your use of technology at work, for social purposes or entertainment, for looking up information, or any other reason.

MODULE 2: USE OF TECHNOLOGY

USE01. In the past 30 days, which have you used to access the Internet? **Check all that apply.**

   a. Personal smartphone (such as iPhone, Blackberry, Android)
   b. Work smartphone
   c. Personal computer
   d. Work computer
   e. Personal iPad or other tablet computer
   f. Work iPad or other tablet computer

74

g. Gaming console (such as Playstation, Xbox)
h. Computers at an Internet café
i. Computers at a school, library, recreation or community center
j. Computer of a friend, relative, or co-worker
k. Other (please specify) _____
l. I have not accessed the Internet in the past 30 days

USE02. Have you downloaded mobile applications, or "apps," to the smartphone or tablet computer you indicated you used to access the Internet? *[Programmer note: Show only to those who selected a, b, e, or f in USE01]*

1 Yes
2 No
3 Not sure

USE03. In the past 30 days, how often did you typically use the following technologies **FOR NON-WORK PURPOSES? Select one response per technology.**

	Every day	4-6 days a week	1-3 days a week	Once every few weeks	I didn't use this technology for non-work purposes
a. Phone calls					
b. Email					
c. Text messaging					
d. Video chat (Skype, VTC)					
e. Social networking sites such as Facebook, Twitter, Google Plus, LinkedIn					
f. Video games (including virtual worlds/social games such as Small Worlds, Second Life, World of Warcraft)					
g. Instant messaging, online chat rooms, forums					
h. Blogs					
i. Other Internet sites					

USE04. In the past 30 days, on the days that you used each technology, about how much time did you spend using it **FOR NON-WORK PURPOSES? Select one response per technology.** *[Programmer note: Exclude items they said they didn't use in USE03]*

	Less than 30 minutes	30-59 min	1-<2 hours	2-<3 hours	3-<4 hours	4+ hours
a. Phone calls						
b. Email						
c. Text messaging						
d. Video chat (Skype, VTC)						
e. Social networking sites such as Facebook, Twitter, Google Plus, LinkedIn						
f. Video games (including virtual worlds/social games such as Small Worlds, Second Life, World of Warcraft)						
g. Instant messaging, online chat rooms, forums						
h. Blogs						
i. Other Internet sites						

USE05. There were one or more technologies that you indicated you did not use in the past 30 days **FOR NON-WORK PURPOSES.** Can you tell us why you don't use the following? **Check all that apply.**
[Programmer note: Include items from USE03 where respondent selected "I don't use this technology"]

	Don't have access	Cost	Not familiar with it	Don't have any interest or need to use it	Have had bad experiences using it in the past	Privacy concerns	Prefer other methods of interaction	Other (Specify)
a. Phone calls								
b. Email								
c. Text messaging								
d. Video chat (Skype, VTC)								
e. Social networking sites such as Facebook, Twitter, Google Plus,								

LinkedIn					
f. Video games (including virtual worlds, social games such as Small Worlds, Second Life, World of Warcraft)					
g. Instant messaging, online chat rooms, forums					
h. Blogs					
i. Other Internet sites					

USE06. In the past 30 days, how often did you typically use the following technologies **AS A PART OF YOUR JOB? Select one response per technology.**

	Every day	4-6 days a week	1-3 days a week	Once every few weeks	I didn't use this technology as a part of my job
a. Phone calls					
b. Email					
c. Text messaging					
d. Video chat (Skype, VTC)					
e. Social networking sites such as Facebook, Twitter, Google Plus, LinkedIn					
f. Video games (including virtual worlds/social games such as Small Worlds, Second Life, World of Warcraft)					
g. Instant messaging, online chat rooms, forums					
h. Blogs					
i. Other Internet sites					

USE07. In the past 30 days, on the days that you used each technology, about how much time each day did you spend using it **AS A PART OF YOUR JOB?**

[Programmer note: Exclude items they said they didn't use in USE06]

	Less than 30 minutes	30-59 min	1-<2 hours	2-<3 hours	3-<4 hours	4+ hours
a. Phone calls						
b. Email						
c. Text messaging						
d. Video chat (Skype, VTC)						
e. Social networking sites such as Facebook, Twitter, Google Plus, LinkedIn						
f. Video games (including virtual worlds/social games such as Small Worlds, Second Life, World of Warcraft)						
g. Instant messaging, online chat rooms, forums						
h. Blogs						
i. Other Internet sites						

USE08. People use technology for lots of different purposes. Please tell us which technologies you use for each of the following purposes. **Check all that apply.**

[Programmer note: Exclude any technologies for which they stated in BOTH USE04 and USE06 "I don't use this technology"]

	Phone calls	Email	Text	Video chat	Social network-ing	Video games	Instant messaging, online chat rooms, forums	Blogs	Other Internet sites	Did not do this
a. Communicate with co-workers										
b. Keep in touch with other Airmen										
c. Keep in touch with friends and family										
d. Keep in touch with acquaintances or										

78

people I don't know very well									
e. Make plans for in-person activities with people									
f. Meet new people									
g. Connect with people I know only online									
h. Complete tasks as part of my Air Force job									
i. Complete tasks as part of a non-military job									
j. Have fun (for entertainment)									
k. Manage/organize my life									
l. Express myself (e.g., journaling, creative outlet)									
m. Find information related to physical health									
n. Find information related to mental health									
o. Connect with others with a similar health concern									
p. Stay informed about Air Force or base news or alerts									
q. Get news, current events									
r. Education, online courses, training									
s. Other (specify)									

MODULE 3: HEALTH INFORMATION

Instructions on screen: Air Force leadership is carefully considering how they can best leverage technology to improve the social and mental fitness of Airmen and their families. The next set of questions asks you about your experiences using technology to find information about mental health for yourself or others, and what you did with that information. There are also some questions about the types of technologies that you would be likely to use to seek information or help for yourself, a loved one, or a fellow Airman should the need arise.

HI01. This first question asks about using the Internet to find mental health information you might need to help yourself or others with emotional or personal problems. Please rate the following statements on a scale from 1 to 5, with a one indicating that you strongly disagree and a five indicating that you strongly agree. **Select one.** [Source: eHEALS (ehealth literacy), modified by RAND to be mental health specific]

	Strongly disagree 1	2	3	4	Strongly agree 5
a. I know how to find helpful mental health resources on the Internet					
b. I know how to use the mental health information I find on the Internet to help me					
c. I have the skills I need to evaluate the mental health resources I find on the Internet					
d. I can tell high quality from low quality mental health resources on the Internet					
e. I feel confident in using information from the Internet to make mental health decisions					

HI02. In the PAST 12 MONTHS, did you use any technologies to learn about mental health topics or resources, including suicide prevention? **Check all that apply.** [Source: RAND]

a. Phone calls
b. Email
c. Text messaging
d. Video chat (Skype, VTC)
e. Social networking sites such as Facebook, Twitter, Google Plus, LinkedIn
f. Video games (including virtual worlds/social games such as Small Worlds, Second Life, World of Warcraft)
g. Instant messaging, online chat rooms, forums
h. Blogs
i. Other Internet sites
j. Other (specify) _____
k. I did not use any technologies for this purpose → SKIP TO HI13

80

HI03. Please list the telephone resources, websites, or other sources where you found information about mental health concerns. **List as many as you can think of.** [Source: RAND]

Websites/location

HI04. In the past 12 MONTHS, did you ever use any technologies to learn about <u>suicide prevention</u> specifically? [Source: RAND, but adapted from National Health Interview Survey]

1 Yes
2 No → SKIP TO HI08

HI05. Was this part of the required Air Force suicide prevention training?

1 Yes
2 No → SKIP TO HI07

HI06. Did you use any other technologies, separate from the Air Force training, to learn about suicide prevention specifically?

1 Yes
2 No → SKIP TO HI08

HI07. Please list the telephone resources, websites, or other sources where you found information about suicide prevention. **List as many as you can think of.** [Source: RAND]

Websites/location

81

HI08. In what ways did you use the mental health information that you found through these technologies? **Check all that apply.** [Source: RAND, but response options adapted from 2007 HTHS-Pew, 2007 C12]

1 I used it to decide whether I should see a mental health professional

2 I talked with a mental health professional about the information I found

3 I used it to make a decision around the best treatment option for me

4 I changed the way I manage my mental health concerns

5 I shared the information with friends or family

6 I shared my thoughts of sadness or loneliness with a friend or peer

7 I shared my thoughts of sadness or loneliness with a health/mental health professional

8 I shared my thoughts about suicide with a friend or peer

9 I shared my thoughts about suicide with a health/mental health professional

10 Other (specify) _____

HI09. In the past 12 MONTHS, has anyone used any technologies to send you information about mental health or mental health care? **Select one.** [Source: RAND]

1 Yes

2 No → SKIP TO HI13

HI10. Did you read, watch or listen to the mental health information that was sent to you?

1 Yes

2 No

3 I don't know

HI11. Did you find that information helpful?

1 Yes

2 No

3 I don't know

HI12. Did that information prompt you to action?

1 Yes

2 No

3 I don't know

Instructions on screen: The previous questions asked you about how you have used technology in the past. Now please think about whether you would be willing to use technology for the prevention or treatment of mental health concerns in the future. In this next set of questions, please select which methods of communication you think would be MOST HELPFUL for the following purposes.

HI13. Which ways would be most helpful for receiving mental health outreach, educational, or screening materials from the Air Force? **Check all that apply.** [Source: RAND]

a. Phone calls
b. Email
c. Text messaging
d. Video chat (Skype, VTC)
e. Social networking sites such as Facebook, Twitter, Google Plus, LinkedIn
f. Online virtual worlds like Second Life, Small Worlds
g. Instant messaging, online chat rooms, forums
h. Blogs
i. Other Internet sites
j. Face to face
k. Other (specify) _____

HI14. Which ways would be most helpful for discussing mental health concerns with others who have similar concerns? **Check all that apply.** [Source: RAND]

a. Phone calls
b. Email
c. Text messaging
d. Video chat (Skype, VTC)
e. Social networking sites such as Facebook, Twitter, Google Plus, LinkedIn
f. Online virtual worlds like Second Life, Small Worlds
g. Instant messaging, online chat rooms, forums
h. Blogs
i. Other Internet sites
j. Face to face
k. Other (specify) _____

HI15. Which ways would be most helpful for seeking advice from a mental health care professional for yourself or others? **Check all that apply.**
[Source: RAND]
a. Phone calls
b. Email
c. Text messaging
d. Video chat (Skype, VTC)
e. Social networking sites such as Facebook, Twitter, Google Plus, LinkedIn
f. Online virtual worlds like Second Life, Small Worlds
g. Instant messaging, online chat rooms, forums
h. Blogs
i. Other Internet sites
j. Face to face
k. Other (specify) _____

HI16. Which ways would be most helpful for receiving mental health care or treatment? **Check all that apply.** [Source: RAND]
a. Phone calls
b. Email
c. Text messaging
d. Video chat (Skype, VTC)
e. Social networking sites such as Facebook, Twitter, Google Plus, LinkedIn
f. Online virtual worlds like Second Life, Small Worlds
g. Instant messaging, online chat rooms, forums
h. Blogs
i. Other Internet sites
j. Face to face
k. Other (specify) _____

MODULE 4: WELL-BEING

Instructions on screen: This next set of questions is designed to help Air Force leadership understand the role technology plays in Airmen's social interactions and also how technology use may be related to Airmen's mental well-being. Remember, you may skip any question you do not want to answer. However, no Air Force leaders or health professionals will have access to your individual responses, and none will receive reports about the responses from their units. Your responses will be used only for research and will be kept completely confidential by the RAND Corporation research team.

[Programmer note: please display the following at the bottom of each page in this Module: **If you experience any distress while taking this survey, we encourage you to contact your unit chain of command, a chaplain, or mental health professional. Support outside of your unit is also available through Military OneSource: www.militaryonesource.mil or 1-800-342-9647. If you or anyone you know is in emotional crisis, you may also contact the Military Crisis Line for confidential help: call 1-800-273-8255 and press 1, or send a text to 838255, or visit the website to initiate a confidential chat: http://veteranscrisisline.net/ActiveDuty.aspx]**

WB01. Please rate the extent to which you agree or disagree with the following statements. **Select one.** [Source: Generalized Problematic Internet Use Scale 2 (GPIUS2)]

	Strongly disagree	Somewhat disagree	Neither agree nor disagree	Somewhat agree	Strongly agree
a. I prefer online social interaction over face-to-face communication.					
b. Online social interaction is more comfortable for me than face-to-face interaction					
c. I prefer communicating with people online rather than face-to-face.					
d. I have used the Internet to talk with others when I was feeling isolated.					
e. I have used the Internet to make myself feel better when I was down.					
f. I have used the Internet to make myself feel better when I've felt upset.					
g. When I haven't been online for some time, I become preoccupied with the thought of going online.					
h. I would feel lost if I was unable to go online.					
i. I think obsessively about going online when I am offline.					
j. I have difficulty controlling the amount of time I spend online.					
k. I find it difficult to control my Internet use.					

l.	When offline, I have a hard time trying to resist the urge to go online.			
m.	My Internet use has made it difficult for me to manage my life.			
n.	I have missed social engagements or activities because of my Internet use.			
o.	I have missed significant amounts of sleep or stayed up all night because of my Internet use.			
o.	My Internet use has created problems for me in my life.			

[Programmer note: please display the following at the bottom of each page in this Module: **If you experience any distress while taking this survey, we encourage you to contact your unit chain of command, a chaplain, or mental health professional. Support outside of your unit is also available through Military OneSource: www.militaryonesource.mil or 1-800-342-9647. If you or anyone you know is in emotional crisis, you may also contact the Military Crisis Line for confidential help: call 1-800-273-8255 and press 1, or send a text to 838255, or visit the website to initiate a confidential chat: http://veteranscrisisline.net/ActiveDuty.aspx]**

WB02. The next questions are about how you feel about different aspects of your life. For each one, please indicate how often you feel that way. **Select one.** [Source: Three-Item Loneliness Scale, Hughes, Waite, Hawkley, and Cacioppo]

	Hardly ever	Some of the time	Often
1. How often do you feel that you lack companionship?			
2. How often do you feel left out?			
3. How often do you feel isolated from others?			

[Programmer note: please display the following at the bottom of each page in this Module: **If you experience any distress while taking this survey, we encourage you to contact your unit chain of command, a chaplain, or mental health professional. Support outside of your unit is also available through Military OneSource: www.militaryonesource.mil or 1-800-342-9647. If you or anyone you know is in emotional crisis, you may also contact the Military Crisis Line for confidential help: call 1-800-273-8255 and press 1, or send a text to 838255, or visit the website to initiate a confidential chat: http://veteranscrisisline.net/ActiveDuty.aspx]**

WB03. Sometimes people use the Internet to harass, bully, "flame," or otherwise attack other people. Please indicate any experience you have had with this behavior in the past 12 months. **Check all that apply.**

1 I have witnessed someone being harassed, bullied, "flamed," or otherwise attacked online.
2 I have been harassed, bullied, "flamed," or otherwise attacked online.
3 I have participated in harassing, bullying, "flaming," or otherwise attacking someone online.
4 I have intervened when I witnessed someone else being harassed, bullied, "flamed," or otherwise attacked online.
5 I have not had any experience with this behavior in the past 12 months.

[*Programmer note*: If any of 1-4 endorsed] We'd like to learn more about these types of activities. Please describe your experience.

WB04. These next three questions ask about your mood over the past two weeks. Please rate how often you have been bothered by either of the following problems in the past TWO WEEKS: **Select one.** As a reminder, your responses to this question will not be reported to your chain of command or trigger any additional action. [Source: PHQ02] [*Programmer note*: *show WB04 a, b, c on one screen*]

a. Little interest or pleasure in doing things
1 Not at all
2 Several days
3 More than half the days
4 Nearly every day

b. Feeling down, depressed, or hopeless
1 Not at all
2 Several days
3 More than half the days
4 Nearly every day

c. If you reported being bothered by the above problems, how difficult have those problems made it for you to do your work, take care of things at home, or get along with other people?
1 Not difficult at all
2 Somewhat difficult
3 Very difficult
4 Extremely difficult

WB05. How would you rate your mental health?
1. Excellent
2. Very good
3. Good
4. Fair
5. Poor

MODULE 5: SOCIAL CAPITAL

Instructions on screen: These last sets of questions ask you about relationships you have with people you know in person, which we'll refer to as knowing "offline," and those you know primarily through online interactions. The first two questions ask about those you know offline; the second two are similar but ask about those you know primarily online. [Source of CAP01-04: Williams' scales for offline/online social capital (2006b)]

CAP01. Please rate the extent to which you agree or disagree with the following statements. Please note, these questions ask about relationships with people you know in person, which we're referring to as knowing "offline." **Select one.**

	Strongly disagree	Somewhat disagree	Neither agree nor disagree	Somewhat agree	Strongly agree
a. There are several people I know offline (in person) I trust to help solve my problems.					
b. There is someone offline I can turn to for advice about making very important decisions.					
c. There is no one offline that I feel comfortable talking to about intimate personal problems.					
d. When I feel lonely, there are several people offline I can talk to.					
e. If I needed an emergency loan of $500, I know someone offline I can turn to.					
f. The people I interact with offline would put their reputation on the line for me.					

88

g. The people I interact with offline would be good job references for me.				
h. The people I interact with offline would share their last dollar with me.				
i. I do not know people offline well enough to get them to do anything important.				
j. The people I interact with offline would help me fight an injustice.				

CAP02. Please rate the extent to which you agree or disagree with the following statements. Please note, these questions ask about relationships with people you know online. By this, we mean people that you met or only know online, or those that you used to know in person but now really only communicate with online. **Select one.**

	Strongly disagree	Somewhat disagree	Neither agree nor disagree	Somewhat agree	Strongly agree
a. There are several people online I trust to help solve my problems.					
b. There is someone online I can turn to for advice about making very important decisions.					
c. There is no one online that I feel comfortable talking to about intimate personal problems.					
d. When I feel lonely, there are several people online I can talk to.					
e. If I needed an emergency loan of $500, I know someone online I can turn to.					
f. The people I interact with online would put their reputation on the line for me.					
g. The people I interact with online would be good job references for me.					
h. The people I interact with online would share their last dollar with me.					
i. I do not know people online well enough to get them to do anything					

important.	
j. The people I interact with online would help me fight an injustice.	

MODULE 1B: FINAL DEMOGRAPHICS, OVERALL TECHNOLOGY ASSESSMENT AND WRITTEN COMMENTS

Instructions on screen: Thank you for taking the time to complete the survey. Before we conclude, we have just a few more questions about your background, and then we'll ask for your overall sense of how these technologies affect you.

DEM09. What is your current relationship status? **Check all that apply.**
1 Never married
2 In a relationship with a boyfriend or girlfriend
3 Married
4 Divorced
5 Separated
6 Widowed

DEM10. Do you have any children under the age of 18? **Select one.**
1 Yes
2 No

DEM11. What is the highest degree or level of school that you have completed? **Select one.**

1 High school diploma or GED
2 Associate degree (includes AA, AS) or vocational/technical diploma (for example, trade school certification)
3 Bachelor's degree or equivalent (includes BA, AB, BS)
4 Graduate degree (includes MA, MS, MBA, PhD, MD, JD, DVM)

DEM12. Are you Spanish/Hispanic/Latino? *[Source: standard DoD race/ethnicity categories]* **Select one.**

1 No, I am not Spanish/Hispanic/Latino
2 Yes, I am Mexican, Mexican-American, Chicano, Puerto Rican, Cuban, or other Spanish/Hispanic/Latino

DEM13. What is your race? Mark one or more races to indicate what you consider yourself to be. *[Source: standard DoD race/ethnicity categories]*

1 White
2 Black or African-American
3 American Indian or Alaska Native
4 Asian (e.g., Asian Indian, Chinese, Filipino, Japanese, Korean, Vietnamese)
5 Native Hawaiian or other Pacific Islander (e.g., Samoan, Guamanian, or Chamorro)
6 Other (specify): _____

OTA01. Which technologies are most important for your social and mental well-being when you are TDY, deployed, or have another temporary geographic relocation away from your home station? **Please check only those you feel are most important to you.**

a. Phone calls
b. Email
c. Text messaging
d. Video chat (Skype, VTC)
e. Social networking sites such as Facebook, Twitter, Google Plus, LinkedIn
f. Video games (including virtual worlds/social games such as Small Worlds, Second Life, World of Warcraft)
g. Instant messaging, online chat rooms, forums
h. Blogs
i. Other Internet sites
j. Other (specify) _____

OTA02. During your most recent deployment, what was the impact of the following technologies on your social and mental well-being? *[Programmer note: Show only to those who said yes to DEM06 and/or DEM07]*

	Mostly positive	Somewhat positive	Neutral	Somewhat negative	Mostly negative	Not available during deployment	Available but I didn't use it
a. Phone calls							
b. Email							
c. Text messaging							

d. Video chat (Skype, VTC)				
e. Social networking sites such as Facebook, Twitter, Google Plus, LinkedIn				
f. Video games (including virtual worlds/social games such as Small Worlds, Second Life, World of Warcraft)				
g. Instant messaging, online chat rooms, forums				
h. Blogs				
i. Other Internet sites				

OTA03. What impact do these technologies have on your overall well-being and life satisfaction today? [***Programmer note:*** *show all technologies, not just ones they say they use personally*]

	Mostly positive	Somewhat positive	Neutral	Somewhat negative	Mostly negative
a. Phone calls					
b. Email					
c. Text messaging					
d. Video chat (Skype, VTC)					
e. Social networking sites such as Facebook, Twitter, Google Plus, LinkedIn					
f. Video games (including virtual worlds/social games such as Small Worlds, Second Life, World of Warcraft)					
g. Instant messaging, online chat rooms, forums					
h. Blogs					
i. Other Internet sites					

CONCLUSION:

Please feel free to let us know anything else you would like to tell us about the use of technology, social networks, the Air Force, and well-being. Note: We will not be able to read your comments in real time. If you are in distress and need help, we encourage you to contact your unit chain of command, a chaplain, or mental health professional. Support outside of your unit is also available through Military OneSource:

92

www.militaryonesource.mil or 1-800-342-9647. If you or anyone you know is in emotional crisis, you may also contact the Military Crisis Line for confidential help: call 1-800-273-8255 and press 1, or send a text to 838255, or visit the website to initiate a confidential chat: http://veteranscrisisline.net/ActiveDuty.aspx *]*

Instructions on screen: Thank you again for your time today completing this survey.

Survey Invitation

MEMORANDUM FOR: Rank/Name   Date

SUBJECT: Technologies to Promote Social and Psychological Well-Being in the Air Force Survey AF12-059SG3

You have been randomly selected by a computer to participate in a survey to help educate Air Force leadership about the role of the Internet, mobile phones, social media, and other technologies in Airmen's lives.

The Air Force Surgeon General, Lt Gen Charles B. Green, has commissioned the RAND Corporation, a nonprofit nonpartisan research company, to conduct a pioneering survey on this topic, with the aim of helping to shape Air Force policy and the way the Air Force uses technology to support the social and mental well-being of Airmen and their families.

Should you choose to participate in this voluntary survey—and we hope you will—we will ask you some questions about what technologies you use, what you use them for, and what you think about them. We will also ask you a few questions about your own social interactions and mental health. Additionally, we will ask you about using technologies to access information and support services to enhance the social and mental fitness of yourself and others.

All of your answers will be completely confidential. No one in the Air Force will know whether you have participated in the study. No Air Force leaders or health professionals will have access to your individual responses, and none will receive reports about the responses from their units.

Your participation is very important and will help ensure that Air Force leadership correctly understands positive and negative aspects of these technologies and takes effective approaches to relevant policy and support service outreach. You may, however, opt out of the study without any negative consequences to your assignments, promotions, or benefits. Also you may answer only the questions you feel comfortable answering or stop taking the survey at any time.

The Air Force survey control number for this officially sanctioned study is:

[number]. You may verify that this survey is on the list of approved surveys by visiting this Air Force web page:
[web link]

Click the link below to view a survey support memo from Air Force Deputy Surgeon General, Major General Thomas W. Travis. This link is accessible only when connected to a .mil network:
[web link]

The survey is available on the Internet and can be accessed via computer or smartphone—you do not have to be on a .mil network to access it. The survey takes approximately 15–20 minutes to complete, depending on your responses. To access the survey, use the following link:
[link to survey web address]

To opt out of the survey click here: [link to opt out notification]
To opt out of future reminders, click here: [link to reminder opt out]

(NOTE: If clicking any of the links in this email does not automatically take you to the correct web page, please copy the address and paste it into your Internet browser.)

If you are unable to access the survey, please contact the survey help desk at [email address and phone number].

The survey will be available until August 25th for active-component Airmen and October 30th for the reserve component.

PLEASE TAKE THE TECHNOLOGIES TO PROMOTE SOCIAL AND PSYCHOLOGICAL WELL-BEING IN THE AIR FORCE SURVEY TODAY!

V/R,
Laura L. Miller, PhD
Senior Social Scientist
RAND Corporation

Sampling and Weighting Methods

This appendix provides technical details about the sampling design and the analytic weights used in this study.

Sampling Design

Population of Interest and Sampling Frame

The target population for the study was all Airmen age 18 and older. Because the survey was going to be administered online and participants were going to be recruited through email, we limited the population to Airmen who had at least one email address (whether official or personal) listed in the Air Force personnel-file extracts from May 2012. Further research would be necessary to determine whether Airmen who have current email addresses on file differ meaningfully from those who do not with regard to the topics in this survey.

We had planned to test for nonresponse bias by conducting a confidential survey (where only the researchers would be able to link respondents to their survey responses) and then destroying the link between responses and identities once our analyses were completed. However, the Air Force Research Oversight and Compliance Office informed us that according to their interpretation of Air Force and DoD policy, if RAND intended to create an identifiable link to the data, survey approval could be granted only under the condition that we permit them access to those data "to ensure [research] subject protection." Concerned about safeguarding Airman responses, we opted instead to conduct an anonymous survey, where not even the researchers can identify who participated. We were, however, able to compare the basic demographics of the Airman population at large with the demographics reported in the survey. To guard against potential bias due to variation in response rates, we weighted the survey data according to key demographic characteristics.

Sampling Design

The sampling design sought to balance analytic precision of the study's results with the Air Force's aim to reduce the frequency with which Airmen are asked to participate in Air Force–sponsored surveys. We wanted to have the ability to analyze survey responses by key subgroups, including Air Force affiliation (active, guard, and reserve), rank group (officer, enlisted), gender (women, men), and age group (18–24, 25–34, 35–44, 45+). More specifically, we wanted to have the ability to analyze rank group, gender, and age group within each of the active-duty, guard, and reserve responses (e.g., guard officers or reserve Airmen ages 35–44). The sample was not designed, however, to obtain precise estimates of and allow powerful comparisons between finer subgroups (e.g., male officers ages 25–34 and female officers ages 25–34).

To balance our goals, we first calculated the minimum random sample size we could obtain and still expect to solicit enough responses to analyze the results by the key subgroups. Using the Air Force Personnel Center's Interactive Demographic Analysis System (IDEAS) in February 2012, we calculated the frequency of Airmen belonging to the cells obtained as the crossing of any two of the variables of interest listed above (affiliation, rank, gender, and age).

The adopted sample design is not proportional to size. In other words, we did not allocate the sample proportional to the size of the three affiliation groups, because that strategy would have called for surveying a greater number of active-duty Airmen than was necessary to meet our analytic goals.

For each of the three Air Force affiliations, we found that a minimum sample of 1,500 Airmen would meet the study's analytical goal, with the single exception of active-duty Airmen ages 45 or older (see Table C.1). From our power calculation, we figured that we needed at least 200 people in each category in order to be able to detect differences of at least 14 percentage points with excellent power (80 percent) between any two groups.[1] Thus 1,500 participants would guarantee that for all but one cell. The total population of active-duty Airmen 45 years old or older is relatively small (N = 11,442), so we oversampled this group to ensure an adequate number of responses. Based on typical response rates for similar surveys of military personnel, we anticipated responses from about one-third of the invited Airmen. Thus, to account for nonresponse, the sample size was increased.

In summary, the adopted sampling design was relatively straightforward: Within every affiliation, we drew a simple random sample of 4,500 Airmen, with the exception of the active-duty group, from which we sampled 437 additional Airmen.

Any Airman for whom email contact information was missing from his or her personnel file was replaced through the same random-sampling process until a complete sample with email information was achieved.

Due to lower-than-expected response rates, to obtain a sufficient number of respondents without moving the survey closing date, we doubled the number of invitations during the survey administration period, drawing additional 4,500-person samples of each component (active duty, guard, and reserve Airmen).

---

[1] We developed this calculation in consultation with colleagues Larry Hanser, Nelson Lim, Lou Mariano, and Al Robbert as a part of an earlier research project.

Table C.1. Expected Demographic Distribution of a Random Sample of Active-Duty, Guard, and Reserve Airmen

	Active	Guard	Reserve
Enlisted	1,208	1,293	1,199
Officer	292	207	301
Men	1,215	1,222	1,117
Women	285	278	383
Age 17–24	485	249	220
Age 25–34	653	517	531
Age 35–44	310	417	405
Age 45+	60	316	345

Weighting Methods

Survey respondents consisted of 1,634 active-duty, 977 guard, and 868 reserve Airmen, for a total of 3,479 Airmen. To achieve the analysis goals, it was important to prepare an analytic sample representative of the Air Force military population with respect to the key subgroups (affiliation, rank group, gender, and age group). Due to differential response patterns and the adopted sampling design—the three affiliation groups are not equally represented in the population—we deemed it necessary to build post-stratification weights so that, for example, the proportion of active-duty women in the sample matches or is close to the proportion of active-duty women in the Air Force military population.

Because the joint distribution of affiliation, rank, gender, and age is known in the population, the computation of an "initial" post-stratification weight was straightforward. The crossing of the four variables defines 48 strata, or cells. For each cell h, we define an "initial" post-stratification weight w_h as the ratio between the population size of the cell (N_h) and the number of respondents in the cell: $w_h = N_h/r_h$ for $h = 1, \ldots, 48$.

Post-stratification weights have the advantage of "improving accuracy of survey estimates both by reducing bias and by increasing precision" (Little, 1993), very much the way a stratified sample would do. However, this advantage is somewhat erased when some cells are small.

This, unfortunately, is our case. In fact, we do not have responses for the cells whose population size is very small—the cells for officers in the youngest age category.

The most used methodology to address the small-cell problem consists of collapsing the small cells with neighboring cells. Neighboring cells are collapsed because they should be more similar to each other than cells that are not neighboring. In particular, response patterns might be similar across neighboring cells. For this reason, we decided to collapse the cell of active-duty female officers aged 18 to 24 with the cell of active-duty female enlisted Airmen aged 18 to 24. More generally, we collapsed the two rank cells defined by crossing affiliation and gender within the 18–24 age category. The rest of the cells were untouched. Collapsing modifies the post-stratification weights for only those cells that are collapsed.

If i and j are the two cells to be collapsed, the post-stratification weight for those subjects that belong to either cell i or j is given by $w_{i,j} = (N_i + N_j) / (r_i + r_j)$. The weights for the other cells h are equal to w_h as defined above.

Collapsing should improve the accuracy of the survey estimates and obtain a less-biased age distribution in the sample.

We weighted our sample of active-duty Airmen to ensure that the proportions in it were demographically representative of the larger Air Force population (N = 328,667). We used the following demographic categories:

- Age group (18–24,[2] 25–34, 35–44, 45 and older)
- Gender (men, women)
- Rank group (enlisted, officer).

The expected distribution of the respondents across these three demographic categories would be that of a random sample of the active-duty component. We therefore weighted the proportions of respondents to those of the entire Air Force active-duty population, using this expected distribution (see Table C.2).

---

[2] We used Air Force population data for 17- to 24-year-old Airmen to estimate the proportion of 18- to 24-year-old Airmen in our sample. For human subjects protection purposes, we excluded 17-year-olds before drawing the random sample of invitees.

Table C.2. Respondent Demographics Used to Weight the Analytic Sample to Be Proportionate to the Air Force Military Population

	Respondents[a]		Weighted Analytic Sample (Percent)
	N	Percent	
Component			
Active	1,634	47	65
Guard	977	28	21
Reserve	868	25	14
Rank			
Enlisted	2,617	75	82
Officer	862	25	18
Gender			
Men	2,741	79	80
Women	738	21	20
Age group			
18–24	354	10	27
25–34	1,121	32	41
35–44	1,056	30	23
45+	948	27	10

[a] Active-duty Airmen ages 45 and higher were oversampled, as were Airmen in the guard and reserve, who are older, on average, than active-duty Airmen.

Military Programs Relating to ICT, Mental Health, and Well-Being

This appendix lists a sample of programs that use ICT to support military mental health and well-being.

- **After Deployment** (afterdeployment.t2.health.mil)

  Afterdeployment.org is a website designed to assist Service members with common post-deployment mental health issues: post-traumatic stress and war memories, conflict at work, depression, anger, sleep problems, alcohol and drug abuse, stress, relationship problems, children and deployment, spiritual guidance and fitness, living with physical injuries, and health and wellness. The site offers information and self-guided exercises for these conditions. Service members can also complete assessments on the site and receive immediate feedback consisting of a recommendation for future action tailored to the assessment results. The website also has a multimedia component with videos of Service members discussing issues they encountered during and after deployment and an interactive component with online workshops.

- **Army G–1 Suicide Prevention** (www.armyg1.army.mil/hr/suicide/default.asp)

  The G1 Suicide Prevention website provides educational material and resources designed to diminish suicide ideation and behavior.

- **Brainline.org**

  Brainline.org is a website that offers information and other content pertaining to traumatic brain injury (TBI), largely through video and written articles. Users can ask TBI experts specific questions and receive answers to these questions by video or written articles. The website also offers webcasts, personal accounts by individuals who have TBI, a calendar of events, research updates, and a resource directory. Materials can be viewed in English or Spanish. In addition to the website itself, brainline.org has a significant social network presence, with a large Facebook group and a large following on Twitter.

- **Military OneSource** (militaryonesource.mil)

  The Military OneSource website provides information on and access to resources in a multitude of areas ranging from mental health to education. Service members and their families can receive nonmedical counseling through the 24-hour call center or online IM for communicating in real time, or Military OneSource can refer them for face-to-face sessions in their community.

- **Military Pathways** (mentalhealthscreening.org/programs/military)
  Military Pathways offers in-person programs, as well as a number of online and telephone programs. Mental health screening tools and educational materials are provided on the Military Pathways website. Military Pathways also provides online screening for a number of different mental health disorders and provides feedback and suggestions based on the results of the screening. Military Pathways offers telephone screening and online resources, including videos, health care resources, information on benefits, and articles.

- **Navy MORE (My Ongoing Recovery Experience)** (navymore.org)
  Navy MORE is a program designed to help individuals recover from substance abuse. The program includes recovery coaches who provide phone support and monitor the recovery process. The program also offers online recovery modules, an online library, and workbook activities. Individuals participating in the program can keep an online journal, track their goals and commitments on an online calendar, and connect online with others in the recovery community.

- **Pdhealth.mil** (www.pdhealth.mil)
  Pdhealth.mil is a website geared toward the families of Service members. It provides information pertaining to mental health through educational materials and media articles.

- **Real Warriors Campaign** (realwarriors.net)
  The Real Warriors Campaign is designed to improve resilience, recovery, and reintegration for Service members. It operates through a number of different mediums, including an interactive website that includes educational materials on mental health, video testimonies, message boards, news updates, and podcasts. The Real Warriors Campaign also has a strong social media presence.

- **Resilience Training (formerly Battlemind)** (www.resilience.army.mil)
  Resilience Training is a program focused on the preparation of Army soldiers, including those in the Army National Guard and Army reserve, for deployment. The program consists largely of life-cycle and deployment-cycle training modules for soldiers, their partners, medical personnel, and military chaplains. Training modules that address National Guard and reserve component-specific issues are also available. Resilience Training has a corresponding website with information and resources specific to the training modules and pertaining to resilience more generally.

- **SimCoach** (simcoach.org)
  SimCoach is an interactive virtual human to which users communicate their mental health concerns. The virtual human gathers relevant mental health information from the users

and provides guidance. This guidance can consist of directing a user to online materials or assisting him or her in seeking live mental health care.

- **Stress Gym**

  Stress Gym is a web-based program created to improve Service members' ability to manage stress. It consists of nine modules completed at a pace directed by the user. The program also allows users to track their stress levels and coping over time. One study of sailors found a significant decline in stress after use of the program, with those who completed more modules showing a greater decrease in stress (Williams et al., 2010).

- **Traumatic Brain Injury: The Journey Home** (traumaticbraininjuryatoz.org)

  Traumatic Brain Injury: The Journey Home is an online program designed to help those suffering from TBI and their families. Its website contains educational content, personal accounts, and multimedia presentations, all pertaining to the needs of those suffering from TBI.

- **HeartMath** (heartmath.com)

  HeartMath provides training to improve the resiliency of Service members. It offers a computer program designed to train both Service members and health professionals effective stress management. HeartMath also provides a handheld device that Service members can use to obtain biofeedback as they practice self-regulation techniques.

- **National Center for Telehealth and Technology** (t2.health.mil)

  The National Center for Telehealth and Technology has developed seven applications for smart devices (phones and tablets): Breathe 2 Relax, Co-Occurring Conditions Toolkit, Mild-TBI Pocket Guide, PTSD Coach, T2 Mood Tracker, Tactical Breather, and LifeArmor. Breathe 2 Relax and Tactical Breather are applications that help individuals manage stress and anxiety through breathing exercises. Co-Occurring Conditions Toolkit is an electronic version of a publication of the same name and includes interactive decision trees. Mild-TBI Pocket Guide is an application designed for health care providers and includes materials to improve the provision of mild-TBI treatment. PTSD Coach is an application designed to help individuals suffering from PTSD; it has features on self-assessment, symptom management, PTSD education, and support finding. T2 Mood Tracker is an application that provides a means for individuals to track their emotions over a period of months. LifeArmor is an application designed to correspond with the afterdeployment.com website, which provides many of the materials it uses and much of the content.

- **TRICARE Assistance Program** (tricare.mil/tamp)

  One aspect of the TRICARE Assistance Program provides nonmedical counseling services to Service members through Skype. The program also provides online chat services 24 hours a day, seven days a week. The counseling services provided through this program are confidential and nonreportable.

- **Virtual Reality and Innovative Technology Applications** (www.itapages.com/home.asp)

  The Virtual Reality and Innovative Technology Applications program emphasizes the development and dissemination of virtual reality and virtual world technologies to address mental health issues. The program's virtual-world endeavors utilize Second Life technologies.

Telemental Health Guidelines

In recent years, many local and professional psychology associations have created guidelines for the proper provision of telemental health. These guidelines expand on the American Psychological Association (APA) Statement on Services by Telephone, Teleconferencing, and Internet (1995), which, though not providing specific guidelines, states that the basic ethical standards for face-to-face services also apply to telemental health services (APA, 2010). Local and professional psychological associations have recently been releasing specific guidelines for telemental health services, and there is a large amount of consensus across many of these organizations.

Licensing

One guideline that all associations agree on is that of licensing. The Canadian Psychological Association (CPA), the Ohio Psychological Association (OPA), the American Mental Health Counselors Association (AMHCA), the American Counseling Association (ACA), the American Telemedicine Association (ATA), and the National Board of Certified Counselors (NBCC) all state in their official codes of ethics or official guidelines that practitioners of telemental health must follow the licensing laws of the jurisdictions where both the practitioner and client reside (ACA, 2005; AMHCA, 2010; CPA, 2006; NBCC, 2005; OPA, 2010). This often requires that practitioners of interstate telemental health be licensed in multiple states. The recently passed National Defense Authorization Act exempts military health care providers, including civilians, from these licensing restrictions by allowing them to provide services with an out-of-state license if they provide services in any DoD-affiliated facility or any facility authorized by the Secretary of Defense (U.S. Congress, 2012). Moreover, guest licenses are offered in many states that allow practitioners to provide services for a short period of time. Practitioners may also receive an Interjurisdictional Practice Certificate, which the Association of State and Provincial Psychology Boards created to facilitate the temporary provision of services outside the jurisdiction of the practitioner (DeAngelis, 2012).

Informed Consent

Another area of consensus across the CPA, the OPA, the AMHCA, and the ACA is that of informed consent. All of these associations state that practitioners of telemental health must obtain informed consent from their clients prior to providing services. According to the organizations' guidelines, practitioners must discuss with their potential clients the specific risks, benefits, and privacy limitations associated with telemental health, along with possible alternatives to telemental health (ACA, 2005; AMHCA, 2010; ATA, 2009; CPA, 2006; NBCC, 2005; OPA, 2010). The guidelines of the OPA (2010) require that psychologists exchange emergency contact information with their clients and that they discuss with their clients a communication policy. The CPA has a similar guideline that requires practitioners of telemental health to discuss their security practices with their clients and reach an agreement with them regarding security (CPA, 2006). The AMHCA and the CPA also instruct practitioners of telemental health to take measures to verify the identity of their clients (AMHCA, 2010; CPA, 2006).

Confidentiality

The official guidelines of the CPA, the OPA, the ATA, and the AMHCA all include precautions that practitioners of telemental health should take to ensure the confidentiality of their clients (AMHCA, 2010; CPA, 2006; OPA, 2008). The OPA instructs psychologists to use secure communication channels, including encryption software, whenever possible (OPA, 2008). The AMHCA addresses the safe transfer of confidential client information in its code of ethics, which instructs psychologists to transfer client information to an authorized third party only when both parties have secure transferring and receiving capabilities (AMHCA, 2010). The official guidelines of the CPA instruct practitioners of telemental health to store all client records and information in a secure manner that ensures client confidentiality (CPA, 2006).

Use of Email

The American Medical Association (AMA) and the Federation of State Medical Boards of the United States (FSMB) have issued guidelines for the use of email in health settings. The AMA's Code of Ethics (American Medical Association, 2003) asserts first that email should not be used to establish new patient relationships, but only to supplement those already established in person. Furthermore, health care providers should ensure the privacy and confidentiality of emails with patients by having a professional password-protected email account that only the providers can access and using email only on a password-protected computer. The Code of Ethics also states that health practitioners should not share patient emails with others unless given explicit permission by the patient. The FSMB (2002) states that all patient email should be part of the patient's medical record, while the AMA states that the physician may choose to enter a summary of the electronic communication into the record.

Quality of Care

The CPA, the OPA, the NBCC, the AMHCA, and the ACA all present guidelines to practitioners of telemental health to ensure the delivery of quality care.

Consistent across all the guidelines is the requirement that practitioners of telemental health services be trained in the provision of these services (ACA, 2005; AMHCA, 2010; CPA, 2006; NBCC, 2005; OPA, 2008). The CPA and the AMHCA extend this guideline by stating that practitioners of telemental health services should not provide services in which they have not demonstrated competence in person or that are outside their area of expertise (AMHCA, 2010; CPA, 2006).

All of the aforementioned associations also state that practitioners of telemental health must ensure that their potential clients are suited emotionally, intellectually, and physically and that they are capable of receiving services (ACA, 2005; AMHCA, 2010; CPA, 2006; NBCC, 2005; OPA, 2008). The CPA and the OPA explicitly state that practitioners must determine the suitability and capability of potential clients for telemental health services during an initial assessment (CPA, 2006; OPA, 2008). An additional guideline set by the CPA and the ACA regarding quality care is that if practitioners of telemental health believe that their clients would benefit more from face-to-face services, they should terminate telemental health services and provide face-to-face services themselves or refer the clients to a practitioner who can provide these services (ACA, 2005; CPA, 2006). The ATA (2009b) provides many practical guidelines for effective telemental health services, including ensuring that the room in which the client is located is sufficiently lit that there are no shadows on his or her face and that the gaze angle is as small as possible.

The ATA also provides a review of evidence-based practices for telemental health (2009b) which states that evidence suggests that telemental health services are effective in a number of domains, including, but not limited to, assessment, psycho-education, and psychotherapy— particularly cognitive behavioral therapy. Computer cognitive behavioral therapy is mentioned in the joint VA and DoD guidelines for the management of major depressive disorder as a viable treatment for service members suffering from mild to moderate depression (Department of Veterans Affairs and Department of Defense, 2009). The joint VA and DoD guidelines for the management of post-traumatic stress also mention telemedicine interventions as treatment for PTSD, although they state that more research on the effectiveness of these interventions is needed before a strong recommendation for their use can be made (Department of Veterans Affairs and Department of Defense, 2010). Both VA and DoD guidelines for management of depression and PTSD suggest the use of telemental health when the provision of face-to-face treatment is not feasible.

Online Suicide Resources and Interventions

This appendix lists a sample of ICT-enabled suicide-information resources and interventions.

- **It Gets Better** (www.itgetsbetter.org)

  Designed as a way to help LGBTQ youth deal with the stress and difficulties of their adolescence, this project posts videos online in which individuals express support for LGBTQ youth and emphasize the notion that their lives will improve. Its website also includes information on resources and a blog. It Gets Better has a large social media presence, with popular Facebook and Twitter accounts and a YouTube site with more than 44,000 subscribers.

- **The Trevor Project** (www.thetrevorproject.org)

  A nonprofit organization focused on suicide prevention among LGBTQ youth, the Trevor Project operates a 24-hour lifeline for LGBTQ youth who wish to talk with trained counselors. The project also provides an online chat service and an asynchronous question-and-answer service, and it hosts a social network for LGBTQ users ages 13–24.

- **International Suicide Prevention Wiki**

  (http://suicideprevention.wikia.com/wiki/International_Suicide_Prevention_Directory)

  An online repository of suicide resources around the world, the International Suicide Prevention Wiki was designed to assist individuals struggling with thoughts of suicide in finding help.

- **I'm Alive** (www.imalive.org)

  I'm Alive is an online network that provides crisis counseling through online chat with trained counselors.

- **Tumblr Suicide Watch** (www.tswatch.tumblr.com)

  Tumblr Suicide Watch is a blog that monitors Tumblr (a popular blog-hosting platform) users. It allows people to post information about "Tumblr users with suicidal intentions" so that others may provide support.

- **Google** (www.google.com)

  Google has set up a feature whereby the phone number of the National Suicide Prevention Lifeline automatically appears at the top of the screen whenever anyone enters the word suicide as the search term. The implementation of this feature, according to

Google, was associated with a 9-percent increase in the amount of calls to the number (Google Unveils New Samaritans Prompt, 2010).

- **Facebook**

Facebook allows users to report the content of others' sites that they deem to be of a suicidal nature. The reporting form specifies:

> IMPORTANT: If you've encountered a direct threat of suicide on Facebook, please contact law enforcement or a suicide hotline immediately. If the person you're worried about is a member of the US military community, be sure to mention this so they can provide this person with custom support. (Facebook, undated a)

The potentially suicidal user who posted the reported content is then sent a message that includes a link for a live chat with a counselor (Weeks, 2011). On its safety resources page, Facebook also provides links to organizational websites that can help users identify and respond to suicidal behavior (Facebook, undated b).

Bibliography

Adamic, Lada, Debra Lauterbach, Chun-Yuen Teng, and Mark S. Ackerman, "Rating Friends Without Making Enemies," *Proceedings of the Fifth International AAAI Conference on Weblogs and Social Media*, 2011.

Ahlstrom, Michelle, Neil R. Lundberg, Ramon Zabriskie, Dennis Eggett, and Gordon B. Lindsay, "Me, My Spouse, and My Avatar," *Journal of Leisure Research*, Vol. 44, No. 1, 2012, pp. 1–22.

Ahuja, Manju K., and John E. Galvin, "Socialization in Virtual Groups," *Journal of Management*, Vol. 29, No. 2, 2003, pp. 161–185.

Alao, Adekola, Maureen Soderberg, Elyssa Pohl, and Abosede L. Alao, "Rapid Communication: Cybersuicide: Review of the Role of the Internet on Suicide," *Cyberpsychology and Behavior*, Vol. 9, No. 4, 2006, pp. 489–493.

American Counseling Association, *ACA Code of Ethics*, Alexandria, VA: American Counseling Association, 2005.

American Medical Association, "The Use of Electronic Mail," *American Medical Code of Ethics,* Opinion 5.026, 2003, in "Ethical Guidelines for Use of Electronic Mail Between Patients and Physicians," web page, http://www.ncbi.nlm.nih.gov/pubmed/14735881. As of March 14, 2014:
http://www.ama-assn.org//ama/pub/physician-resources/medical-ethics/code-medical-ethics/opinion5026.page

———, *Emotional and Behavioral Effects of Video Games and Internet Overuse*, AMA House of Delegates 2007 Annual Meeting, Council on Science and Public Health Report 12-A-07, 2007.

———, *Guidelines for Patient-Physician Electronic Mail*, Chicago, IL: American Medical Association, 2000.

American Mental Health Counselors Association, *Principles for AMHCA Code of Ethics,* Alexandria, Va., 2010.

American Psychiatric Association, *Diagnostic and Statistical Manual of Mental Disorders*, 5th ed., Arlington, Va.: American Psychiatric Publishing, 2013.

American Psychological Association, *Ethical Principles of Psychologists and Code of Conduct,* Washington, D.C., 2010.

Anderson, Janna Q., and Lee Rainie, "Millenials Will Benefit and Suffer Due to Their Hyperconnected Lives," *Pew Internet*, Washington, D.C.: Pew Research Center, 2012.

Andersson, Gerhard, Jan Bergstrom, Fredrik Hollandare, Per Carlbring, Viktor Kaldo, and Lisa Ekselius, "Internet-based Self-help for Depression: Randomised Controlled Trial," *British Journal of Psychiatry*, Vol. 187, 2005, pp. 456–461.

Andersson, Gerhard, and Pim Culjpers, "Internet-Based and Other Computerized Psychological Treatments for Adult Depression: A Meta-Analysis," *Cognitive Behaviour Therapy*, Vol. 38, No. 4, 2009, pp. 196–205.

Andreassen, Cecilie S., Torbjorn Torsheim, Geir S. Brunborg, and Stale Pallesen, "Development of a Facebook Addiction Scale," *Psychological Reports*, Vol. 110, No. 2, 2012, pp. 501–517.

Baker, Laurence, et al., "Use of the Internet and E-mail for Health Care Information," *Journal of the American Medical Association*, Vol. 289, No.18, 2003, pp. 2400–2406.

Barak, Azy, "Emotional Support and Suicide Prevention Through the Internet: A Field Project Report," *Computers in Human Behavior*, Vol. 23, No. 2, 2007, pp. 971–984.

Barak, Azy, Meyran Boniel-Nissim, and John Suler, "Fostering Empowerment in Online Support Groups," *Computers in Human Behavior*, Vol. 24, 2008, pp. 1867–1883.

Barak, Azy, Liat Hen, Meyran Boniel-Nissim, and Na'ama Shapira, "A Comprehensive Review and a Meta-Analysis of the Effectiveness of Internet-Based Psychotherapeutic Interventions," *Journal of Technology in Human Services*, Vol. 26, No. 2-4, 2008, pp. 109–160.

Barak, Azy, and Ofra Miron, "Writing Characteristics of Suicidal People on the Internet: A Psychological Investigation of Emerging Social Environments," *Suicide and Life-Threatening Behavior*, Vol. 35, No. 5, 2005, pp. 507–524.

Baume, Pierre, Christopher H. Cantor, and Andrew Rolfe, "Cybersuicide: The Role of Interactive Suicide Notes on the Internet," *Crisis*, Vol. 18, No. 2, 1997, pp. 73–79.

Baym, Nancy, *Personal Connections in the Digital Age*, Malden, Mass.: Polity Press, 2010.

Beard, Keith W., "Internet Addiction: Current Status and Implications for Employees," *Journal of Employment Counseling*, Vol. 39, 2002, pp. 2–11.

Becker, Katja, and Martin H. Schmidt, "When Kids Seek Help On-line: Internet Chat Rooms and Suicide," *Reclaiming Children and Youth*, Vol. 13, No. 4, 2005, pp. 229–230.

Bessiere, Katherine, Sara Kiesler, Robert Kraut, and Bonka S. Boneva, "Effects of Internet Use and Social Resources on Changes in Depression," *Information, Communication and Society*, Vol. 11, No. 1, 2008, pp. 47–70.

Bessiere, Katherine, Sarah Pressman, Sara Kiesler, and Robert Kraut, "Effects of Internet Use on Health and Depression: A Longitudinal Study," *Journal of Medical Internet Research*, Vol. 12, No. 1, 2010, p. e6.

Biddle, Lucy, Jenny Donovan, Keith Hawton, Navneet Kapur, and David Gunnell, "Suicide and the Internet," *Public Health,* Vol. 336, 2008, pp. 800–802.

Blumberg, Stephen J., and Julian V. Luke, *Wireless Substitution: Early Release of Estimates from the National Health Interview Survey, January–June 2013,* Atlanta, Ga.: Centers for Disease Control and Prevention, 2013.

Boellstorff, Tom, *Coming of Age in Second Life: An Anthropologist Explores the Virtually Human*, Princeton, N.J.: Princeton University Press, 2008.

boyd, danah,[1] Jenny Ryan, and Alex Leavitt, "Pro-Self-Harm and the Visibility of Youth-Generated Problematic Content," *I/S: A Journal of Law and Policy for the Information Society,* Vol. 7, No. 1, 2010, pp 1–32.

Brenner, Viktor, "Psychology of Computer Use: XLVII. Parameters of Internet Use, Abuse and Addiction: The First 90 Days of the Internet Usage Survey," *Psychological Reports*, Vol. 80, 1997, pp. 879–882.

Brewin, Chris, Bernice Andrews, and John D. Valentine, "Meta-Analysis of Risk Factors for Posttraumatic Stress Disorder in Trauma-Exposed Adults," *Journal of Consulting and Clinical Psychology*, Vol. 68, No. 5, 2000, pp. 748–766.

Brodie, Mollyann, et al., "Health information, the Internet, and the Digital Divide," *Health Affairs*, Vol. 19, No. 6, 2000, pp. 255–265.

Canadian Psychological Association, *Providing Psychological Services via Electronic Media*, Ottowa, Ontario: Canadian Psychological Association, 2006. As of July, 29, 2012: http://www.cpa.ca/aboutcpa/committees/ethics/psychserviceselectronically/

Caplan, Scott E., "A Social Skill Account of Problematic Internet Use," *Journal of Communication,* December 2005, pp. 721–736.

———, "Preference for Online Social Interaction: A Theory of Problematic Internet Use and Psychosocial Well-Being," *Communication Research,* Vol. 30, 2003, pp. 625–648.

———, "Relations Among Loneliness, Social Anxiety, and Problematic Internet Use," *Cyberpsychology and Behavior,* 2007, Vol. 10, No. 2, pp. 234–242.

———, "Theory and Measurement of Generalized Problematic Internet Use: A Two-Step Approach," *Computers in Human Behavior*, Vol. 26, 2010, pp. 1089–1097.

---

[1] Note that danah boyd's legal name is spelled with all lower-case letters.

———, "Problematic Internet Use and Psychosocial Well-being: Development of a Theory-based Cognitive-Behavioral Measurement Instrument," *Computers in Human Behavior*, Vol. 18, 2002, pp. 553–575.

Caplan, Scott, Dmitri Williams, and Nick Yee, "Problematic Internet Use and Psychosocial Well-Being Among MMO Players," *Computers in Human Behavior*, Vol. 25, No. 6, 2009, pp. 1312–1319.

Carr, Nicholas, *The Shallows: What the Internet is Doing to Our Brains,* New York: W. W. Norton and Company, 2011.

Chen, Yi-Fan, and James E. Katz, "Extending Family to School Life: College Students' Use of the Mobile Phone," *International Journal of Human-Computer Studies*, Vol. 67, 2008, pp. 179–191.

Chou, Chien, Linda Condron, and John C. Belland, "A Review of the Research on Internet Addiction," *Educational Psychology Review*, Vol. 17, No. 4, 2005, pp. 363–388.

Corliss, Jonathan, "Introduction: The Social Science Study of Video Games," *Games and Culture*, Vol. 6, No. 1, 2011, pp. 3–16.

Cotten, Shelia R., Melinda Goldner, Timothy M. Hale, and Patricia Drentea, "The Importance of Type, Amount, and Timing of Internet Use for Understanding Psychological Distress," *Social Science Quarterly,* Vol. 92, No. 1, 2011, pp. 119–139.

Couch, Danielle, and Pranee Liamputtong, "Online Dating and Mating: Perceptions of Risk and Health Among Online Users," *Health, Risk and Society*, Vol. 9, No. 3, 2007, pp. 275–294.

Davis, R. A., "A Cognitive-Behavioral Model of Pathological Internet Use," *Computers in Human Behavior,* Vol. 17, 2001, pp. 187–195.

Dean, Elizabeth, Sarah Cook, Joe Murphy, and Michael Keating, "The Effectiveness of Survey Recruitment Methods in Second Life," *Social Science Computer Review*, Vol. 30, No. 3, 2012, pp. 324–338.

DeAngelis, Tori, "Practicing Distance Therapy, Legally and Ethically," *Monitor on Psychology*, Vol. 43, No. 3, 2012, p. 52.

Department of Veterans Affairs and Department of Defense, *Clinical Practice Guideline: Management of Major Depressive Disorder (MDD)*, Washington, DC: Department of Veterans Affairs and Department of Defense, May, 2009.

———, *VA/DoD Clinical Practice Guideline for the Management of Post-Traumatic Stress,* Washington, DC: Department of Veterans Affairs and Department of Defense, October, 2010.

Douglas, Alecia, C., Juline E. Mills, Mamadou Niang, Svetlana Stepchenkova, Sookeun Byun, Celestino Ruffini, Seul Ki Lee, Jihad Loutfi, Jung-Kook Lee, Mikhail Atallah, and Marina Blanton, "Internet Addiction: Meta-synthesis of Qualitative Research for the Decade 1996–2006," *Computers in Human Behavior*, Vol. 24, 2008, pp. 3027–3044.

Duggan, Maeve and Aaron Smith, *Social Media Update 2013*, Pew Research Center, December 2013. As of July 1, 2014:
http://pewinternet.org/Reports/2013/Social-Media-Update.aspx

Eastin, Matthew S., "The Influence of Competitive and Cooperative Group Game Play on State Hostility," *Human Communication Research*, Vol. 33, 2007, pp. 450–466.

Ellison, Nicole B., Charles Steinfield, and Cliff Lampe, "The Benefits of Facebook 'Friends': Social Capital and College Students' Use of Online Social Network Sites," *Journal of Computer-Mediated Communication*, Vol. 12, 2007, pp. 1143–1168.

Entertainment Software Association, *Essential Facts About the Computer and Video Game Industry*, 2013.

Eysenbach, G., J. Powell, O. Kuss, and E. R. Sa, "Empirical Studies Assessing the Quality of Health Information for Consumers on the World Wide Web," *Journal of the American Medical Association*, Vol. 287, No. 20, 2002, pp. 2691–2700.

Facebook, "Report Suicidal Content," web page, undated a. As of July 10, 2014:
http://www.facebook.com/help/contact/?id=305410456169423

———, "Where Can I Find Resources for Identifying and Helping a Friend Who May Be Suicidal?" web page, undated b. As of July 10, 2014:
http://www.facebook.com/help/224061994364693

Federation of State Medical Boards of the United States, Inc., *Model Guidelines for the Appropriate Use of Internet in Medical Practice*, Euless, Tex., 2002.

Fischer, Claude S., *Still Connected: Family and Friends in America Since 1970*, New York: Russell Sage Foundation, 2011.

Fleishman, John A., and Samuel H. Zuvekas, "Global Self-Rated Mental Health: Associations with Other Mental Health Measures and with Role Functioning," *Medical Care*, Vol. 45, No. 7, 2007, pp. 602–609.

Forest, Amanda L., and Joanne V. Wood, "When Social Networking Is Not Working: Individuals with Low Self-esteem Recognize but Do Not Reap the Benefits of Self-disclosure on Facebook," *Psychological Science*, Vol. X, 2012, pp. 1–8.

Fox, Jesse, and Jeremy N. Bailenson, "Virtual Self-Modeling: The Effects of Vicarious Reinforcement and Identification on Exercise Behaviors," *Media Psychology*, Vol. 12, 2009, pp. 1–25.

Fox, Susannah, "The Social Life of Health Information, 2011," *Pew Internet*, Washington, D.C.: Pew Research Center, 2011.

Fox, Susannah, and Maeve Duggan, "Health Online 2013," *Pew Internet*, Washington, D.C.: Pew Research Center, 2013.

Freberg, Karen, Rebecca Adams, Karen McGaughey, and Laura Freberg, "The Rich Get Richer: Online and Offline Social Connectivity Predicts Subjective Loneliness," *Media Psychology Review*, Vol. 3, No. 1, 2010.

Gajadhar, B. J., Y. deKort, W. Ljsselsteijn, and K. Poels, "Where Everybody Knows your Game: The Appeal and Function of Game Cafes in Western Europe," *Ace*, October 29–31, 2009, pp. 28–35.

Gallup, "How Does Gallup Polling Work?" undated. As of June 25, 2014:
http://www.gallup.com/poll/101872/how-does-gallup-polling-work.aspx

Geekaphone, 2011. As of February 3, 2013:
http://www.techfever.net/2011/08/geekaphone-releases-an-infographic-of-the-mobile-gaming-market/

Gentile, Douglas A., Craig A. Anderson, Shintaro Yukawa, Nobuko Ihori, Muniba Saleem, Lin Kam Ming, Akiko Shibuya, Albert Liau, Angeline Khoo, Brad J. Bushman, L. Rowell Huesmann, and Akira Sakamoto, "The Effects of Prosocial Video Games on Prosocial Behaviors: International Evidence from Correlational, Longitudinal, and Experimental Studies," *Prosocial Video Games and Behavior*, Vol. 35, No. 6, 2009, pp. 752–763.

"Google Unveils New Samaritans Prompt," *The Herald (Glasgow)*, November 12, 2010.

Gray, Jennifer B., and Neal D. Gray, "The Web of Internet Dependency: Search Results for the Mental Health Professional," *International Journal of Mental Health Addiction*, Vol. 4, 2006, pp. 307–318.

Greene, Talya, Joshua Buckman, Christopher Dandeker, and Neil Greenberg, "How Communication with Families Can Both Help and Hinder Service Members' Mental Health and Occupational Effectiveness on Deployment," *Military Medicine*, Vol. 175, 2010, pp. 745–749.

Greidanus, Elaine, and Robin D. Everall, "Helper Therapy in an Online Suicide Prevention Community," *British Journal of Guidance & Counseling*, Vol. 38, No. 2, 2010, pp. 191–204.

Gros, Daniel F, Matthew Yoder, Peter W. Tuerk, Brian Lozano, and Ron Acierno, "Exposure Therapy for PTSD Delivered to Veterans via Telehealth: Predictors of Treatment Completion and Outcome and Comparison to Treatment Delivered in Person," *Behavior Therapy*, Vol. 42, No. 2, 2011, pp. 276–283.

GSS: General Social Survey, undated. As of June 25, 2014:
http://www3.norc.org/GSS+Website/

Haas, Ann, Bethany Koestner, Jill Rosenberg, David Moore, Steven J. Garlow, Jan Sedway, Linda Nicholas, Herbert Hendin, John Mann, and Charles B. Nemeroff, "An Interactive Web-Based Method of Outreach to College Students at Risk for Suicide," *Journal of American College Health*, Vol. 57, No. 1, 2008, pp. 15–22.

Hampton, Keith N., "Grieving for a Lost Network: Collective Action in a Wired Suburb," The *Information Society*, Vol. 19, 2003, pp. 417-428.

Hampton, Keith N., Lauren S. Goulet, Lee Rainie, and Kristen Purcell, "Social Networking Sites and Our Lives: How People's Trust, Personal Relationships, and Civic and Political Involvement Are Connected to Their Use of Social Networking Sites and Other Technologies," *Pew Internet*, Washington, D.C.: Pew Research Center, 2011.

Hampton, Keith N., Oren Livio, and Lauren Sessions, "The Social Life of Wireless Urban Spaces: Internet Use, Social Networks, and the Public Realm," paper presented at the pre-conference workshop at the International Communication Association (ICA) Conference, Chicago, Ill., May 20–21, 2009.

Harris, Keith M., John P. McLean, and Jeanie Sheffield, "Examining Suicide-Risk Individuals Who Go Online for Suicide-related Purposes," *Archives of Suicide Research,* Vol. 13, No. 3, 2009, pp. 264–276.

Hawi, Nazir, "Internet Addiction Among Adolescents in Lebanon," *Computers in Human Behavior*, Vol. 28, 2012, pp. 1044–1053.

Hilty, Donald M., Daphne C. Ferrer, Michelle Burke Parish, Barb Johnston, Edward J. Callahan, and Peter M. Yellowlees, "The Effectiveness of Telemental Health: A 2013 Review," *Telemedicine and e-Health,* Vol. 19, No. 6, 2013, pp. 444–454.

Hinduja, Sameer, and Justin W. Patchin, "Bullying, Cyberbullying, and Suicide," *Archives of Suicide Research*, Vol. 14, No. 3, 2010, pp. 206–221.

Hitsch, Gunter J., Ali Hortacsu, and Dan Ariely, "What Makes You Click: An Empirical Analysis of Online Dating," *Society for Economic Dynamics*, Meeting Paper 207, 2005.

Hoge, C. W., J. L. Auchterlonie, and C. S. Milliken, "Mental Health Problems, Use of Mental Health Services, and Attrition From Military Service After Returning from Deployment to Iraq or Afghanistan," *Journal of the American Medical Association,* Vol. 295, No. 9, March 1, 2006, pp. 1023–1032.

Hughes, Mary Elizabeth, Linda J. Waite, Louise C. Hawkley, and John T. Cacioppo, "A Short Scale for Measuring Loneliness in Large Surveys: Results from Two Population-Based Studies," *Research on Aging*, Vol. 26, No. 6, 2004, pp. 655–672.

Institute of Medicine, *Treatment for Posttraumatic Stress Disorder in Military and Veteran Populations*, Washington, D.C.: The National Academies Press, 2012.

Ito, Mizuko, Heather Horst, Judd Antin, Megan Finn, Arthur Law, Annie Manion, Sarai Mitnick, David Schlossberg, and Sarita Yardi, *Hanging Out, Messing Around, and Geeking Out: Kids Living and Learning with New Media*, Cambridge, Mass.: MIT Press, 2010.

Ivory, James D., and Sriram Kalyanaraman, "The Effects of Technological Advancement and Violent Content in Video Games on Players' Feelings of Presence, Involvement, Physiological Arousal, and Aggression," *Journal of Communication*, Vol. 57, 2007, pp. 537–555.

Jones, Norman, Rachel Seddon, Nicola T. Fear, Pete McAllister, Simon Wessely, and Neil Greenberg, "Leadership, Cohesion, Morale, and the Mental Health of UK Armed Foces in Afghanistan, " *Psychiatry,* Vol. 75, No. 1, 2012.

Jong, Michael, "Managing Suicides via Videoconferencing in a Remote Northern Community in Canada," *International Journal of Circumpolar Health*, Vol. 63, No. 4, 2004, pp. 422–428.

Kobayashi, Tetsuro, "Bridging Social Capital in Online Communities: Heterogeneity and Social Tolerance of Online Game Players in Japan," *Human Communication Research*, Vol. 32, 2010, pp. 546–569.

Kraut, Robert, Sara Kiesler, Bonka Boneva, Jonathon Cummings, and Vicki Helgeson, "Internet Paradox Revisited," Human-Computer Interaction Institute, Paper 101, 2001.

Kraut, Robert, Michael Patterson, Vicki Lundmark, Sara Kiesler, Tridas Mukopadhyay, and William Scherlis, "Internet Paradox: A Social Technology That Reduces Social Involvement and Psychological Well-being," *American Psychologist*, Vol. 53, No. 9, 1998, pp. 1017–1031.

Kroenke, Kurt, Robert L. Spitzer, and Janet B. W. Williams, "The Patient Health Questionnaire-2: Validity of a Two-Item Depression Screener," *Medical Care*, Vol. 41, No. 11, 2003, pp. 1284–1292.

Krysinska, Karolina, and Diego DeLeo, "Telecommunication and Suicide Prevention: Hopes and Challenges for the New Century," *Omega*, Vol. 55, No. 3, 2007, pp. 237–253.

Kuhn, S., A. Romanowski, C. Schilling, et al., "The Neural Basis of Video Gaming," *Translational Psychiatry*, Vol. 1, 2011, pp. 1–5.

Lapierre, C. B., A. F. Schwegler, and B. J. LaBauve, "Posttraumatic Stress and Depression Symptoms in Soldiers Returning from Combat Operations in Iraq and Afghanistan," *Journal of Trauma and Stress,* Vol. 20, No. 6, December 2007, pp. 933–943.

Lee, Ook, and Mincheol Shin, "Addictive Consumption of Avatars in Cyberspace," *Cyberpsychology & Behavior*, Vol. 7, No. 4, 2004, pp. 417–420.

Lenhart, Amanda, "Adults and Social Network Websites," *Pew Internet*, Washington, D.C.: Pew Research Center, Pew Internet Project Data Memo, January 14, 2009.

Lenhart, Amanda, Joseph Kahne, Ellen Middaugh, Alexandra R. Macgill, Chris Evans, and Jessica Vitak, "Teens, Video Games, and Civics," *Pew Internet*, Washington, D.C.: Pew Research Center, 2008.

Lenhart, Amanda, Kristen Purcell, Aaron Smith, and Kathryn Zickuhr, "Social Media and Mobile Internet Use among Teens and Young Adults," *Pew Internet*, Washington, D.C.: Pew Research Center, 2010.

Little, R. J. A., "Post-Stratification: A Modeler's Perspective," *Journal of the American Statistical Association*, Vol. 88, 1993, pp. 1001–1012.

Madden, Mary, and Kathryn Zickuhr, "65% of Online Adults Use Social Networking Sites," *Pew Internet*, Washington, D.C.: Pew Research Center, 2011.

Malaby, Thomas, *Making Virtual Worlds*, Ithaca, N.Y.: Cornell University Press, 2009.

Marche, Stephen, "Is Facebook Making Us Lonely?" *The Atlantic*, May 2012, pp. 1–10.

Mark, Gloria J., Stephen Voida, and Armand V. Cardello, "'A Pace Not Dictated by Electrons': An Empirical Study of Work Without Email," *CHI 2012: Proceedings of the SIGCHI Conference on Human Factors in Computing Systems*, NewYork: Association for Computing Machinery, 2012.

Markel, Howard, "The D.S.M. Gets Addiction Right," *New York Times*, June 5, 2012.

Martin, C. B., "Routine Screening and Referrals for PTSD After Returning from Operation Iraqi Freedom in 2005, U.S. Armed Forces," *Medical Surveillance Monthly Report*, Vol. 14, No. 6, September/October 2007, pp. 2–7.

Mawani, Farah N., and Heather Gilmour, "Validation of Self-Rated Mental Health," *Health Reports*, Vol. 21, No. 3, 2010, pp. 61–75.

McGene, Juliana, *Social Fitness and Resilience: A Review of Relevant Constructs, Measures, and Links to Well-Being*, Santa Monica, Calif.: RAND Corporation, RR-108-AF, 2013. As of July 1, 2014:
http://www.rand.org/pubs/research_reports/RR108.html

McGonigal, Jane, *Reality Is Broken: Why Games Make Us Better and How They Can Change the World*, London: Penguin Press, 2011.

McKenna, Katelyn Y. A., Amie S. Green, and Marci E. J. Gleason, "Relationship Formation on the Internet: What's the Big Attraction?" *Journal of Social Issues,* Vol. 58, No. 1, 2002, pp.9–31.

Meerkerk, Gert-Jan, Regina Van Den Eijnden, and Henk Garretsen, "Predicting Compulsive Internet Use: It's All About Sex!" *Cyberpsychology and Behavior,* Vol. 9, No. 1, 2006, pp. 95–103.

Mehlum, Lars, "The Internet, Suicide, and Suicide Prevention," *Crisis*, Vol. 21, No. 4, 2000, pp. 186–188.

Mesch, Gustavo S., and Ilan Talmud, "Internet Connectivity, Community Participation, and Place Attachment: A Longitudinal Study," *American Behavioral Scientist*, Vol. 53, No. 8, 2010, pp. 1095–1110.

Miller, Geoffrey, "The Smartphone Psychology Manifesto," *Perspectives on Psychological Science*, Vol. 7, No. 3, 2012, pp. 221–237.

Miller, John K., and Kenneth J. Gergen, "Life on the Line: The Thereapeutic Potentials of Computer-Mediated Conversation," *Journal of Marital and Family Therapy*, Vol. 24, No. 2, 1998, pp. 189–202.

Miranda, Christine, *Fear of Missing Out*, New York: JWT, 2011.

Mitchell, Kimberly J., Kathryn A. Becker-Blease, and David Finkelhor, "Inventory of Problematic Internet Experiences Encountered in Clinical Practice," *Professional Psychology: Research and Practice*, Vol. 36, No. 5, 2005, pp. 498–509.

Mitchell, Kimberly J., David Finkelhor, and Kathryn A. Becker-Blease, "Classification of Adults with Problematic Internet Experiences: Linking Internet and Conventional Problems from a Clinical Perspective," *Cyberpsychology & Behavior*, Vol. 10, No. 3, 2007, pp. 381–392.

Mitchell, Mary M., Michael Shayne Gallaway, Amy M. Millikan, and Michael Bell, "Interaction of Combat Exposure and Unit Cohesion in Predicting Suicide-Related Ideation Among Post-Deployment Soldiers," *Suicide and Life-Threatening Behavior,* Vol. 42, No. 5, 2012, pp. 486–494.

Morahan-Martin, Janet, and Phyllis Schumacher, "Incidence and Correlates of Pathological Internet Use Among College Students," *Computers in Human Behavior*, Vol. 16, 2000, pp. 13–29.

———, "Loneliness and Social Uses of the Internet," *Computers in Human Behavior*, Vol. 19, 2003, pp. 659–671.

Moreno, Megan A., Lauren A. Jelenchick, Katie G. Egan, Elizabeth Cox, Henry Young, Kerry E. Gannon, and Tara Becker, "Feeling Bad on Facebook: Depression Disclosures by College Students on a Social Networking Site," *Depression and Anxiety*, Vol. 28, 2011, pp. 447–455.

Muise, Amy, Emily Christofodes, and Serge Desmarais, "More Information Than You Ever Wanted: Does Facebook Bring Out the Green-Eyed Monster of Jealousy?" *Cyberpsychology & Behavior*, Vol. 12, No. 4, 2009, pp. 441–444.

Nardi, Bonnie, *My Life as a Night Elf Priest: An Anthropological Account of World of Warcraft*, Ann Arbor, Mich.: University of Michigan Press, 2010.

National Board of Certified Counselors, *Code of Ethics,* Greensboro, N.C., 2005.

New Jersey Department of Consumer Affairs, *Internet Dating Safety Act*, N.J.S.A. 56:8-168 et seq., undated.

Newark-French, Charles, "Mobile App Usage Further Dominates Web, Spurred by Facebook," *Flurry,* blog, January 9, 2012. As of July 1, 2014: http://www.flurry.com/bid/80241/Mobile-App-Usage-Further-Dominates-Web-Spurred-by-Facebook#.U7Q4KY1dVOE

Ng, Brian D., and Peter Wiemer-Hastings, "Addiction to the Internet and Online Gaming," *Cyberpsychology & Behavior*, Vol. 8, No. 2, 2005, pp. 110–113.

Nielsen, "Friends & Frenemies: Why We Add and Remove Facebook Friends," Newswire, December 19, 2011. As of July 1, 2014: http://www.nielsen.com/us/en/insights/news/2011/friends-frenemies-why-we-add-and-remove-facebook-friends.html

Niemz, Katie, Mark Griffiths, and Phil Banyard, "Prevalence of Pathological Internet Use Among University Students and Correlations with Self-Esteem, the General Health Questionnaire (GHQ), and Distribution," *Cyberpsychology and Behavior*, Vol. 8, No. 6, 2005, pp. 562–569.

Norman, Cameron D., and Harvey A. Skinner, "eHEALS: The eHealth Literacy Scale," *Journal of Medical Internet Research*, Vol. 8, No. 4, 2006a, p. e27.

———, "eHealth Literacy: Essential Skills for Consumer Health in a Networked World," *Journal of Medical Internet Research*, Vol. 8, No. 2, 2006b, p. e9.

Ohio Psychological Association, *Telepsychology Guidelines,* Columbus, Ohio, 2010.

Ozer, Emily J., Suzanne R. Best, Tami L. Lipsey, and Daniel S. Weiss, "Predictors of Posttraumatic Stress Disorder and Symptoms in Adults: A Meta-Analysis," *Psychological Trauma: Theory, Research, Practice and Policy*, Special Volume, No. 1, 2008, pp. 3–36.

Patchin, Justin, and Sameer Hinduja, "Cyberbullying and Self-Esteem," *Journal of School Health*, Vol. 80, No. 12, 2010, pp. 614–621.

Pennebaker, J. W., S. D. Barger, and J. Tiebout, "Disclosure of Traumas and Health among Holocaust Survivors," *Psychosomatic Medicine*, Vol. 51, No. 5, 1989, pp. 577–589.

Pew Research Internet Project, *The Web at 25 in the U.S.*, Washington, D.C.: Pew Research Center, February 2014. As of July 1, 2014:
http://www.pewinternet.org/2014/02/25/the-web-at-25-in-the-u-s

Przybylski, Andrew K., Netta Weinstein, Kou Muruyama, Martin Lynch, and Richard Ryan, "The Ideal Self at Play: The Appeal of Video Games That Let You Be All You Can Be," *Psychological Science,* Vol. X, 2011, pp. 1–8.

Quan-Haase, Anabel, Barry Wellman, and James Witte, "Capitalizing on the Net: Social Contact Civic Engagement, and Sense of Community," University of Toronto, Canada, manuscript, 2002.

Rosenstiel, Tom, Amy Mitchell, Kristen Purcell, and Lee Rainie, *How People Learn About Their Local Community*, Washington, D.C.: Pew Research Center, 2011.

Russell, Daniel W., "UCLA Loneliness Scale (Version 3): Reliability, Validity, and Factor Structure," *Journal of Personality Assessment*, Vol. 66, No. 1, 1996, pp. 20–40.

Russell, Dan, Letitia A. Peplau, and Carolyn E. Cutrona, "The Revised UCLA Loneliness Scale: Concurrent and Discriminant Validity Evidence," *Journal of Personality and Social Psychology*, Vol. 39, 1980, pp. 472–480.

Ryan, Richard M., and Edward L. Deci, "Self-Determination Theory and the Facilitation of Intrinsic Motivation, Social Development, and Well-Being," *American Psychologist*, Vol. 55, No. 1, 2000, pp. 68–78.

Saleem, Muniba, Craig A. Anderson, and Douglas Gentile, "Effects of Prosocial, Neutral, and Violent Video Games on College Students' Affect," *Aggressive Behavior*, Vol. 38, No. 4, July/August 2012, pp. 1–9.

Schrock, Andrew, "Examining Social Media Usage: Technology Clusters and Social Network Site Membership," *First Monday,* Vol. 14, No. 1-5, January 2008.

Schroeder, Ralph, *Being There Together: Social Interaction in Shared Virtual Environments*, New York: Oxford University Press, 2010.

Seay, A. Fleming, and Robert E. Kraut, "Project Massive: Self-Regulation and Problematic Use of Online Gaming," *CHI '07: Proceedings of the SIGCHI conference on Human Factors in Computing Systems*, New York: Association for Computing Machinery, 2007, pp. 829–838.

Second Life, "Your World, Your Imagination," web page, undated. As of June 29, 2014:
www.secondlife.com

Sessions, Lauren F., "How Offline Gatherings Affect Online Communities," *Information, Communication, & Society,* Vol. 13, No. 3, 2010, pp. 375–395.

Shapira, Nathan A., Mary C. Lessig, Toby D. Goldsmith, Steven T. Szabo, Martin Lazoritz, Mark S. Gold, and Dan J. Stein, "Problematic Internet Use: Proposed Classification and Diagnostic Criteria," *Depression and Anxiety*, Vol. 17, 2003, pp. 207–216.

Sheldon, Pavica, "The Relationship Between Unwillingness-to-Communicate and Students' Facebook Use," *Journal of Media Psychology*, Vol. 20, No. 2, 2008, pp. 67–75.

Shiovitz-Ezra, Sharon, and Liat Ayalon, "Use of Direct Versus Indirect Approaches to Measure Loneliness in Later Life," *Research on Aging*, Vol. 34, No. 5, 2012, pp. 572–591.

Shirky, Clay, *Cognitive Surplus: How Technology Makes Consumers into Collaborators*, London: Penguin Press, 2010.

Silenzio, Vincent M., Paul R. Duberstein, Wan Tang, Naiji Lu, Xin Tu, and Christopher M. Homan, "Connecting the Invisible Dots: Reaching Lesbian, Gay, and Bisexual Adolescents and Young Adults at Risk for Suicide Through Online Social Networks," *Social Science & Medicine*, Vol. 69, 2009, pp. 469–474.

Smith, Aaron, "35% of American Adults Own a Smartphone," *Pew Internet*, Washington, D.C.: Pew Research Center, 2011.

———, "Americans and Text Messaging," *Pew Internet*, Washington, D.C.: Pew Research Center, 2011.

———, "Mobile Access 2010," *Pew Internet*, Washington, D.C.: Pew Research Center, 2010.

Smith, Aaron, Lee Rainie, and Kathryn Zickuhr, "College Students and Technology," *Pew Internet*, Washington, D.C.: Pew Research Center, 2011.

Smith, T. C., M. A. K. Ryan, D. L. Wingard, D. J. Slymen, J. F. Sallis, D. Kritz-Silverstein, and Team for the Millennium Cohort Study, "New Onset and Persistent Symptoms of Post-Traumatic Stress Disorder Self Reported After Deployment and Combat Exposures: Prospective Population Based US Military Cohort Study," *British Medical Journal,* January 15, 2008.

Song, Indeok, Robert Larose, Matthew S. Eastin, and Carolyn A. Lin, "Internet Gratifications and Internet Addiction: On the Uses and Abuses of New Media," *Cyberpsychology & Behavior*, Vol. 7, No. 4, 2004, pp. 384–394.

Spitzer, Robert L., et al., "Validity and Utility of the PRIME-MD Patient Health Questionnaire in Assessment of 3000 Obstetric-Gynecologic Patients: The PRIME-MD Patient Health

Questionnaire Obstetrics-Gynecology Study," *American Journal of Obstetrics and Gynecology*, Vol. 183, No. 3, 2000, p. 759.

Spitzer, Robert L., Kurt Kroenke, and Janet B. W. Williams, "Validation and Utility of a Self-Report Version of PRIME-MD," *Journal of the American Medical Association*, Vol. 282, No.1, 1999, pp. 1737–1744.

Steinfield, Charles, Joan M. DiMicco, Nicole B. Ellison, and Cliff Lampe, "Bowling Online: Social Networking and Social Capital Within the Organization," *C&T '09: Proceedings of the Fourth International Conference on Communities and Technologies*, New York: Association for Computing Machinery, 2009, pp. 245–254.

Steinfield, Charles, Nicole B. Ellison, and Cliff Lampe, "Social Capital, Self-Esteem, and Use of Online Social Network Sites: A Longitudinal Analysis," *Journal of Applied Developmental Psychology*, Vol. 29, 2008, pp. 434–445.

Steinkuehler, Constance, and Dmitri Williams, "Where Everybody Knows Your (Screen) Name: Online Games as 'Third Places,'" *Journal of Computer-Mediated Communication*, Vol. 11, 2006, pp. 885–909.

Stone, Deborah M., Catherine W. Barger, and Lloyd Potter, "Public Health Trianing Online: The National Center for Suicide Prevention Training," *American Journal of Preventive Medicine*, Vol. 29, 2005, pp. 247–251.

Sussman, Stephanie W., and Lee Sproull, "Straight Talk: Delivering Bad News Through Electronic Communication," *Information Systems Research*, Vol. 10, No. 2, 1999, pp. 150–166.

Tamir, Diana, and Jason P. Mitchell, "Disclosing Information About the Self Is Intrinsically Rewarding," Proceedings of the National Academy of Sciences of the United States of America, *PNAS Early Edition*, 2012, pp. 1–6.

Tanielian, Terri L., and Lisa Jaycox (eds.), *Invisible Wounds of War: Psychological and Cognitive Injuries, Their Consequences, and Services to Assist Recovery,* Santa Monica, Calif.: RAND Corporation, MG-720-CCF, 2008. As of July 1, 2014: http://www.rand.org/pubs/monographs/MG720.html

Thom, Katey, Gareth Edwards, Ivana Nakarada-Kordic, Brian McKenna, Anthony O'Brien, and Raymond Nairn, "Suicide Online: Portrayal of Website-Related Suicide by the New Zealand Media," *New Media and Society*, Vol. 13, No. 8, 2011, pp. 1355–1372.

Toma, Catalina, Jeffrey Hancock, and Nicole Ellison, "Separating Fact from Fiction: An Examination of Deceptive Self-Presentation in Online Dating Profiles," *Personality and Social Psychology Bulletin*, Vol. 34, No. 8, 2008, pp. 1023–1036.

Tufekci, Zeynep, "Grooming, Gossip, Facebook, and MySpace," *Information, Communication, and Society,* Vol. 11, No. 4, 2008, pp. 544–564.

Turkle, Sherry, *Alone Together: Why We Expect More from Technology and Less from Each Other,* New York: Basic Books, 2011.

U.S. Congress, *National Defense Authorization Act,* H.R. 1540, Washington, D.C., 2012.

Vincent, Jane, "Emotion and the Mobile Phone," University of Surrey, Digital World Resarch Centre Faculty of Arts and Human Sciences, manuscript, 2009.

———, "Emotional Attachment and Mobile Phones," *Knowledge, Technology, and Policy*, Vol. 19, No. 1, 2006, pp. 39–44.

Vitak, Jessica, Nicole B. Ellison, and Charles Steinfield, "The Ties That Bind: Re-examining the Relationship Between Facebook Use and Bonding Social Capital," *Proceedings of the 44th Hawaii International Conference on System Sciences*, Kauai, Hawaii, 2011.

Walther, Joseph B., Geri Gay, and Jeffrey T. Hancock, "How Do Communication and Technology Researchers Study the Internet?" *Journal of Communication*, September 2005, pp. 632–657.

Warschauer, Mark, *Technology and Social Inclusion: Rethinking the Digital Divide*, Cambridge, Mass.: MIT Press, 2003.

Weeks, William B., Amy E. Wallace, Alan N. West, Hilda R. Heady, and Kara Hawthorne, "Research on Rural Veterans: An Analysis of the Literature," *The Journal of Rural Health,* Vol. 24, No. 4, 2008, pp. 337–344.

Whang, Leo Sang-Min, Sujin Lee, and Geunyoung Chang, "Internet Over-Users' Psychological Profiles: A Behavior Sampling Analysis on Internet Addiction," *Cyberpsychology and Behavior,* Vol. 6, No. 2, 2003, pp. 143–150.

Widyanto, Laura, and Mark Griffiths, "'Internet Addiction': A Critical Review," *International Journal of Mental Health Addiction,* Vol. 4, 2006, pp. 31–52.

Williams, Arthur, Bonnie M. Hagerty, Stever J. Brasington, Joseph B. Clem, and David A. Williams, "Stress Gym: Feasibility of Deploying a Web-Enhanced Behavioral Self-Management Program for Stress in a Military Setting," *Military Medicine,* Vol. 175, No. 7, 2010, pp. 487–493.

Williams, Dmitri, "On and Off the 'Net': Scales for Social Capital in an Online Era," *Journal of Computer-Mediated Communication*, Vol. 11, No. 2, 2006.

Williams, Dmitri, Nicolas Ducheneaut, Li Xiong, Yuanyuan Zhang, Nick Yee, and Eric Nickell, "From Tree House to Barracks: The Social Life of Guilds in World of Warcraft," *Games and Culture*, Vol. 1, No. 4, 2006, pp. 338361.

Williams, Dmitri, Nick Yee, and Scott E. Caplan, "Who Plays, How Much, and Why? Debunking the Stereotypical Gamer Profile," *Journal of Computer-Mediated Communication*, Vol. 13, 2008, pp. 993–1018.

Wilson, Robert E., Samuel D. Gosling, and Lindsay T. Graham, "A Review of Facebook Research in the Social Sciences," *Perspectives on Psychological Science*, Vol. 7, No. 3, 2012, pp. 203–220.

Wilson, Samuel M., and Leighton C. Peterson, "The Anthropology of Online Communities," *Annual Review of Anthropology*, Vol. 31, 2002, pp. 449–467.

Wood, Richard T., "Problems with the Concept of Video Game 'Addiction': Some Case Study Examples," *International Journal of Mental Health Addiction*, Vol. 6, 2008, pp. 169–178.

Yardi, Sarita, and danah boyd, "Tweeting from the Town Square: Measuring Geographic Local Networks," *Proceedings of the International Conference on Weblogs and Social Media*, Vancouver, British Columbia, 2007.

Young, Kimberly S., "Internet Addiction: A New Clinical Phenomenon and Its Consequences," *American Behavioral Scientist*, Vol. 48, No. 4, 2004, pp. 402–415.

———, "Internet Addiction: The Emergence of a New Clinical Disorder," paper presented at the 104th annual meeting of the American Psychological Association, Toronto, Canada, August 15, 1996.

Yuen, C. N., and Michael J. Lavin, "Internet Dependence in the Collegiate Population: The Role of Shyness," *Cyberpsychology & Behavior*, Vol. 7, No. 4, 2004, pp. 379–383.

Zaphiris, Panaviotis, and Chee Siang Ang, *Social Computing and Virtual Communities*, Boca Raton, Fla.: Chapman and Hall/CRC, 2009.

Zickuhr, Kathryn, *Generations 2010*, Washington, D.C.: Pew Research Center, December 2010.